ARTIFICIAL
INTELLIGENCE
IN MEDICINE

ARTIFICIAL INTELLIGENCE IN MEDICINE

From Ethical, Social, and Legal Perspectives

Edited by

JOSEPH J.Y. SUNG
Lee Kong Chian School of Medicine, Nanyang
Technological University, Singapore, Singapore

CAMERON STEWART
Sydney Law School, University of Sydney, Sydney, NSW,
Australia

ELSEVIER

ACADEMIC PRESS
An imprint of Elsevier

Academic Press is an imprint of Elsevier
125 London Wall, London EC2Y 5AS, United Kingdom
525 B Street, Suite 1650, San Diego, CA 92101, United States
50 Hampshire Street, 5th Floor, Cambridge, MA 02139, United States

Notices
Knowledge and best practice in this field are constantly changing. As new research and experience broaden our understanding, changes in research methods, professional practices, or medical treatment may become necessary.

Practitioners and researchers must always rely on their own experience and knowledge in evaluating and using any information, methods, compounds, or experiments described herein. In using such information or methods they should be mindful of their own safety and the safety of others, including parties for whom they have a professional responsibility.

To the fullest extent of the law, neither the Publisher nor the authors, contributors, or editors, assume any liability for any injury and/or damage to persons or property as a matter of products liability, negligence or otherwise, or from any use or operation of any methods, products, instructions, or ideas contained in the material herein.

ISBN 978-0-323-95068-8

For information on all Academic Press publications
visit our website at https://www.elsevier.com/books-and-journals

Publisher: Stacy Masucci
Acquisitions Editor: Linda Buschman
Editorial Project Manager: Barbara L. Makinster
Production Project Manager: Omer Mukthar
Cover Designer: Matthew Limbert

Typeset by STRAIVE, India

Working together
to grow libraries in
developing countries

www.elsevier.com • www.bookaid.org

Contents

Contributors   *ix*

Preface   *xi*

1. Artificial intelligence and the future of medicine   **1**

Joseph J.Y. Sung

Introduction   1

What are the capabilities of AI in 2023?   2

Image-based analysis and diagnosis   2

Clinical decision support and application of precision medicine   3

Prediction of health conditions and treatment outcome   4

Improving accessibility of healthcare and empowering patients with lifestyle

    modification and compliance to treatment   5

Drug and diagnostic test discovery   5

ChatGPT   6

Successful implementation of AI in medicine   8

Obstacles to overcome for implementation of AI in medicine   8

Engaging the clinicians and healthcare providers   9

Successful implementation of AI in medicine   10

References   12

2. Data access, data bias, data equity   **13**

Dennis L. Shung and Loren Laine

Introduction   13

Definition of terms   14

Data access   15

Data bias   18

Data equity   21

Algorithmic stewardship as a framework for mitigation strategy against

    bias and inequity   23

References   24

3. Respect for persons   **27**

Tamra Lysaght, Vicki Xafis, and Cameron Stewart

Introduction	27
A values-based deliberative balancing approach to ethical decision-making	29
Case study: Applying the Framework to AI-assisted PDSS	37
Conclusion	42
References	42

4. Privacy and confidentiality   **45**

Stephen Kai Yi Wong

Introduction	45
Personal health information (PHI)	46
Privacy	46
Confidentiality	47
Privacy and confidentiality in medical and health services	47
Privacy crossing paths with confidentiality	48
Exemptions from privacy breaches	50
Data protection principles (DPP)	50
Accountability	53
Consent	54
Enhanced privacy rights	55
Data ethics	56
AI and digital data ethics in med-health sciences and services	56
From a privacy structure to a privacy culture	57
References	58

5. Black box medicine   **59**

Irwin King, Helen Meng, and Thomas Y.T. Lam

Introduction	59
The three stages of AI development with five principles	60
Data collection	60
Transparency	61
Privacy of data	62
Models and algorithm development	63
Black box medicine: What is acceptable?	68
References	68

6. Clinical evidence   71
Kendall Ho, Sarah Park, Michael Lai, and Simon Krakovsky

Introduction	71
What is AI in clinical medicine?	71
Case examples	73
What is needed for AI?	79
Barriers of AI	81
Future of AI	83
Conclusion	83
References	84

7. Medical AI and tort liability   89
I. Glenn Cohen, Andrew Slottje, and Sara Gerke

Introduction	89
Liability for medical AI	91
Malpractice liability and the standard of care	94
Regulation and preemption as alternative to the common law of torts	98
Conclusion	101
Acknowledgments	101
References	102

8. Regulation of AI in healthcare   105
Colin Gavaghan

Introduction	105
What do we mean by "regulation"?	106
What should be regulated?	107
Premarket regulation	111
Postmarket regulation	111
Transparency and consent	113
Devices or practitioners?	114
References	116

9. Health inequalities in AI machine learning   119
Roger Yat-Nork Chung and Ben Freedman

Introduction	119
Biases in AI machine learning	119
Impact of AI on social determinants of health	125
Conclusion	128
Acknowledgment	129
References	129

10. Human-machine interaction: AI-assisted medicine, instead of AI-driven medicine   131

René F. Kizilcec, Dennis L. Shung, and Joseph J.Y. Sung

Introduction   131
Implementation is the challenge   133
Human-machine interaction   134
AI-CDSSs as clinical team members   136
Conclusion   138
References   139

Index   *141*

Contributors

Roger Yat-Nork Chung
JC School of Public Health and Primary Care; CUHK Centre for Bioethics; CUHK Institute of Health Equity, The Chinese University of Hong Kong, Sha Tin, Hong Kong SAR

I. Glenn Cohen
James A. Attwood and Leslie Williams Professor of Law; Petrie-Flom Center for Health Law Policy, Biotechnology & Bioethics, Harvard Law School, Cambridge, MA, United States

Ben Freedman
Heart Research Institute; Charles Perkins Centre and Faculty of Medicine and Health, University of Sydney, Sydney, NSW, Australia

Colin Gavaghan
School of Law/Bristol Digital Futures Institute, University of Bristol, England, United Kingdom

Sara Gerke
Penn State Dickinson Law, Carlisle, PA, United States

Kendall Ho
University of British Columbia Faculty of Medicine, Emergency Medicine, Vancouver, BC, Canada

Irwin King
Department of Computer Science and Engineering, The Chinese University of Hong Kong, Hong Kong Special Administrative Region

René F. Kizilcec
Department of Information Science, Cornell University, Ithaca, NY, United States

Simon Krakovsky
University of British Columbia, Vancouver, BC, Canada

Michael Lai
University of British Columbia, Vancouver, BC, Canada

Loren Laine
Yale School of Medicine, New Haven; Veterans Administration Connecticut Healthcare System, West Haven, CT, United States

Thomas Y.T. Lam
The Nethersole School of Nursing, The Chinese University of Hong Kong, Hong Kong Special Administrative Region

Tamra Lysaght
Faculty of Medicine and Health, Sydney School of Public Health, Sydney Health Ethics, University of Sydney, Sydney, NSW, Australia

Helen Meng
Department of System Engineering and Engineering Management, The Chinese University of Hong Kong, Hong Kong Special Administrative Region

Sarah Park
University of British Columbia Faculty of Medicine, Emergency Medicine, Vancouver, BC, Canada

Dennis L. Shung
Yale School of Medicine, New Haven, CT, United States

Andrew Slottje
J.D., Class of 2023, Harvard Law School, Cambridge, MA, United States

Cameron Stewart
Sydney Law School, University of Sydney, Sydney, NSW, Australia

Joseph J.Y. Sung
Lee Kong Chian School of Medicine, Nanyang Technological University, Singapore, Singapore

Stephen Kai Yi Wong
Gilt Chambers, 8/F Far East Finance Centre, Central, Hong Kong

Vicki Xafis
University of Melbourne, Melbourne, VIC, Australia

Preface

Many times in life, projects are born out of happy accidents, and that is the case with this book. In 2019, we met each other by chance at a graduation ceremony at the School of Law, University of Sydney, where Joseph was attending the academic procession for the graduation of his godson and Cameron was presiding over the ceremony as acting dean. At that meeting, we came to share a mutual interest in exploring the legal and ethical issues of artificial intelligence (AI) in healthcare. After working together for a time, Joseph proposed the idea that we should bring together experts and get them to write an edited volume—the very volume that you now hold in your hands.

This volume combines the expertise of scholars from around the globe. Each chapter is an attempt to tackle important issues that are currently being faced by health practitioners and patients as AI starts to penetrate deeper into the practice of medicine and allied healthcare. Our main task was to encourage our contributors to tackle a discrete topic of interest and then relate that to broader themes so that the reader will be able to pick up discrete solutions to problems, but in a way that they can also see the bigger picture. We thank all our contributors for their passion and energy, and we are very proud of each chapter in the volume.

We also thank the team from Elsevier, especially Barbara Makinster, who guided us along. Edited volumes always take a long time to emerge, but this volume would probably not have emerged at all without the help of Max Cheung, Manager (Research and External Affairs) at the Lee Kong Chian School of Medicine, Nanyang Technological University, Singapore. Thank you Max for your dedication and hard work.

Joseph and Cameron would like to thank Ben Freeman for introducing them to each other. Joseph would also like to personally thank his wife, Rebecca, for her patience and understanding. Cameron would like to personally thank his wife, Nerida, for her support over many years.

The disruptive technologies of AI are changing medical practice in a rapid and revolutionary way. What is written here is going to be obsolete in a few years, maybe even a few months. However, this book should trigger some thoughts and discussion on the ethical and social issues of applying AI to life-and-death decisions and management. We hope that the book will be used as a resource for laypeople and for future scholarship on AI in health-care. We encourage readers to continue to follow the work of the authors in this volume.

As the WHO Guidance on Ethics and Governance of AI for Health said "If employed wisely, AI has the potential to empower patients and communities to assume control of their own health care better... But if we do not take appropriate measures, AI could also lead to situations where decisions that should be made by providers and patients are transferred to machine, which would undermine human autonomy."

Joseph J.Y. Sung
Cameron Stewart

CHAPTER 1

Artificial intelligence and the future of medicine

Joseph J.Y. Sung
Lee Kong Chian School of Medicine, Nanyang Technological University, Singapore, Singapore

Introduction

The tsunami of artificial intelligence (AI) has arrived in medicine, penetrating every clinical specialty. Deep learning algorithms enable highly sensitive and specific diagnosis of diabetic retinopathy. Breast cancer screening using mammography can be performed by machine-learning devices, saving much time for radiologists. Automated classification of skin conditions using convoluted neural network (CNN) program in smart phones enables dermatologists to make vital diagnosis from a distance. Neural network algorithms can detect and characterize polyps during colonoscopy, reducing the chance of missing such potentially malignant lesions. AI may also facilitate physician-care manager-patient partnership in educating patients, checking compliance, and enhancing self-management in chronic diseases. Increasingly, machine learning devices can replace time-consuming, labor-intensive, repetitive, and mundane tasks of clinicians.

AI comprises any digital system that mimics human reasoning capabilities, including image pattern recognition, language processing, and abstract reasoning and planning. It includes the concept of machine learning, where machines can learn from experience in ways that mimic human behavior, but with the ability to assimilate much more data and with potential for greater accuracy and speed. Machine learning is a research field that has seen recent advances due to exponential increases in computing power, algorithmic coding that mimics the human cognitive process (deep learning), and access to large and linked sources of big data.

The scope of AI can be specific, performing narrowly defined tasks (narrow AI) such as image interpretation, or more general, applying knowledge and skills in different contexts (general AI) such as making a diagnosis and predicting disease outcome. On the other hand, machine learning can also be designated "supervised," in which a dataset is provided for the algorithm

Artificial Intelligence in Medicine
https://doi.org/10.1016/B978-0-323-95068-8.00001-7

to evaluate its performance, or "unsupervised," in which the machine is allowed to extract unknown potential features in developing an algorithm.

The ultimate aim for applying AI into medicine is for improvement of patient experience (quality of care and satisfaction), achieving better clinical outcome (fewer clinic attendance, shorter hospital stay, faster recovery, and lower mortality), and delivering better care at a sustainable cost (reduce per capita cost).

The introduction of AI into medicine may cause disruption of the norms and conventional clinical practice. The concept of "creative destruction" used by Dr. Eric Topol points to the fact that "introduction of radical innovation into the traditional systems becomes the real force for sustained long-term growth. Such growth only comes by destroying the value of established systems and enterprise." So, this is nothing less than a revolution [1]. The question is, how to introduce revolutionary changes and, at the same time, preserve the essentials and the good practice in medicine.

Medicine is not just a science, but it is also an art, a humanistic endeavor that takes care of the most important aspects of life and death. In this book, we focus on not just the science and technology of computer engineering and algorithm, but also the social, ethical, and legal aspects of implementing AI in medicine.

What are the capabilities of AI in 2023?

Five major areas that AI holds promises to healthcare delivery: (1) image-analysis and diagnostics, (2) clinical decision support in applying precision medicine, (3) prediction of health conditions and treatment outcome, (4) improving accessibility of healthcare and empowering patients with lifestyle modification and compliance to treatment, and (5) drug and diagnostic test discovery.

Image-based analysis and diagnosis

Analyzing medical images is a natural fit for "deep learning." Comparing the use of AI data interpretation, AI-assisted image interpretation has made further advances in the diagnosis and prognostication of diseases. The use of AI in radiology holds great promise in improving quality and efficiency in diagnosing various medical conditions. Diego Ardila has shown that with 3D deep learning on low-dose chest computed tomography; lung cancer can be diagnosed with higher accuracy [2]. Stand-alone artificial intelligence for breast cancer detection in mammography has been shown no worse than the average performance of 101 radiologists in breast cancer detection [3]. Equally important to recognize is that it has the power to supplant or at least augment the capabilities of radiologist when focused on certain aspects of image analysis and diagnosis. However, the application of AI in real-life radiology is more complicated. There are many other aspects of diagnostic workflow that radiologist do may not be replaced by AI, at least at this stage. This includes selecting the imaging modalities that should be taken, contributing to treatment plan according to finding in the picture, applying image-guided tissue biopsy, radiological interventional procedures for the treatment of diseases in the blood vessels, tumor, and drainage of abscess or fluid cavity in the body, just to name a few. Therefore, Geoffrey Hinton's calling radiologists "coyote already over the edge of the cliff who hasn't yet looked down" is an overpessimistic view.

Image interpretation can reach far more than radiology. Recognizing the histological features of normal versus pathological tissue has been a cornerstone in medicine. Histopathologists provide diagnosis and classification of disease and suggest prognosis based on the classical features of pathology. AI methods expand, extract quantitative information from digital histopathology images, and recognize features that conventional pathology may not notice [4]. Other image interpretation application also includes use of retinal image for the diagnosis of diabetic retinopathy, retinopathy of prematurity, and age-related macular degeneration [5]. Scientists from Stanford University demonstrated that, using convolutional neural network (CNN), AI is comparable to dermatologist in the classification and recognition of skin cancer [6]. Application of AI in endoscopic imaging has improved the diagnosis of malignant and premalignant condition, detection of polyps, development of objective scoring systems for risk stratification, predicting progression of liver fibrosis and cancer development, and determining which patient with inflammatory bowel disease will benefit from biologic therapy [7]. The application is expanding every day and permeate into different specialties.

Clinical decision support and application of precision medicine

Precision medicine is an emerging approach for prediction of health events, prevention of disease and adverse outcome, tailoring treatment for individuals, as well as predicting the outcome of such measurements, taking into account variability in genes, environment, and lifestyle for each person. This is in contrast to historical one-size-fit-all approach in treatment and prevention of diseases which were developed over centuries for average person with less consideration for the difference between individuals.

To be able to achieve this, it requires three things: (1) collection of huge volume of genomic, environmental, and lifestyle data of high quality and representative of the respective population, (2) immense analytic capability to build models and algorithms for the study of diseases based on strong and weak signals in the individual genome and environment that he/she is living with, and (3) a trustworthy system of data collection, analysis, and utilization by the subjects and the healthcare providers. With the myriad of data, machine learning can analyze, recognize pattern, develop algorithm for diagnosis, and propose treatment. Precision medicine identifies phenotypes of patients with unique healthcare needs. AI technology generates insights, enables the system to reason and learn, and empowers clinician decision-making through augmented intelligence. If this can be achieved, AI will be at the center of medical solutions that are able to personalize health and medical services down to the level of each patient and consumer. Precision medicine will be driven by AI to not just identify the right treatment for a particular patient but also individualize all aspects of healthcare for that patient, including disease risk rating and prognosticate outcome of treatment [8].

Prediction of health conditions and treatment outcome

Medical treatment and hospital stay are adding to healthcare cost exponentially. Every year, millions of patients are being admitted to hospitals, or even intensive care unit, unnecessarily. Either they do not need to be staying in hospital or their length of stay can be significantly shortened. In others, hospitalization and treatment in the ICUs do not improve their outcome. Worldwide, there is a wide variance in the average length of stay for hospitals within a region or nation, and it has been estimated that as many as 20% of these hospital days are avoidable. Research conducted in various countries shows that a substantial proportion of hospital days are not needed or wasted. By providing greater insights into the care path for each patient, through the prediction of clinical outcome by AI, our healthcare system may reduce average length of stay, avoid unnecessary spending on tests and investigations, and reduce the risk of exposing patients to hospital-acquired infections and other adverse events [9].

Improving accessibility of healthcare and empowering patients with lifestyle modification and compliance to treatment

On a population level, AI has promised to benefit not just patients as individuals but also enhance improving the health of a population. This could be achieved by (1) improving the accessibility of health services to places where care delivery is problematic, (2) empowering the patients (or even health subjects) to take care of their health, (3) preventing disease, and (4) changing their lifestyle and compliance to medication.

With the continuous development of cloud technology and AI-assisted diagnosis and treatment, this new way of healthcare delivery will help to improve access to service in underserved communities and countries of lower income. AI will become increasingly pervasive in its application to improve access to health and medical services where they used to be difficult to reach. There are already examples of AI-assisted ultrasound service to help saving life in servicing developing countries around the world. Diagnosis of eye and skin conditions from afar using AI-image interpretation will revolutionize the care of ophthalmological and dermatological conditions. There are many other examples such as the screening of chest radiographs for tuberculosis, ultrasound for pregnant women to assess neonatal development, and more.

On the other hand, AI device may empower patients and healthy subjects to take better care of themselves without coming to clinics and hospitals. AI-assisted rehabilitation tools have been developed to provide program for elderly subjects, patients with chronic illnesses such as stroke, cardiovascular diseases, and cognitive impairment, to exercise their limbs and their mind at home. These programs are often tailored to meet individual's needs and to monitor their progress over time. In order for this AI tools to be implemented and used at home and by the lay public, it has to be user-friendly, affordable and show effectiveness in achieving what it meant to achieve. Additional areas of AI tools include modification of diet, rehabilitation from smoking and drinking habit, encouraging exercise, and active lifestyle, just to name a few. This could create huge impact on population health to prevent diseases and protect health of the aging population.

Drug and diagnostic test discovery

Discovering and bringing new drugs to market is notoriously slow and costly. This is because human biology is complex involving not just genomics, transcriptomic, and metabolomic but also environmental factors and lifestyle (smoking, drinking, and exercise). Despite the human genome project that has been accomplished with successful deciphering of DNA, the progress of using these genome data in finding drugs for the treatment of over 7000 known rare disease is progressing very slowly.

It is against this background that drug researchers and developers are turning to machine learning to make the hunt for new pharmaceuticals quicker, cheaper, and more effective. Deep learning (DL) which uses artificial neural networks (ANNs) open a new path to study compound property, predict activity and reaction in human body and, combined with chemo-informatics and retrosynthetic analysis, may speed up the process of drug development.

There are many other potential applications of AI in medicine, from interpreting genomic and metagenomic data and combining these data with lifestyle and environmental factors, which could make significant contributions in the development of identifying biomarkers (biological and digital) of diseases and individual health to pharma and medical technology development. Potential use of AI in medicine is summarized in Fig. 1.1.

ChatGPT

In November 2022, the launch of ChatGPT has shocked the world of the capability of AI in many fields, including medicine. The arrival of ChatGPT,

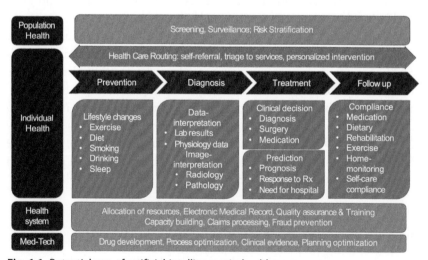

Fig. 1.1 Potential use of artificial intelligence in healthcare.

the latest version of Large Language Model (LLM), has been described as another "iPhone moment." Rowan Curran, an analyst, describes that just as the iPhone and App Store launched by Apple has put the entire computing experience in people's pockets; ChatGPT appears to put the healthcare industry on the cusp of a similar game-changing moment from consultation to referral of cases and perhaps in future the delivery of services.

There are already ad hoc cases showing that ChatGPT is able to make clinical diagnosis of individual patients based on clinical features and test results. By absorbing and integrating knowledge in cyberspace and keeping tracks with millions of scientific publication in the medical literature, ChatGPT can author, or at least co-author, scientific paper, and it can even help candidates to pass professional examination.

Looking forward, it is entirely possible that ChatGPT can be trained to prepare clinical notes from doctors who dictate their findings while seeing patients. The report can be structures to presenting symptoms, ongoing complaints, frequency and severity of attacks, and many other details. That would save many hours of documentation in medical notes.

In future, more advanced LLM can interact with patients directly to ask about more details of the patient's symptom. At home, the medical chatbot function of ChatGPT can help our patients to understand their problem and refer them to doctors more efficiently. How would that improve or create confusion in medical consultation and health service would need time to unveil the development.

There are, at least for today, certain limitations of ChatGPT in medical practice. Currently, the existing ChatGPT algorithm does not cite references and quote evidence to substantiate their claims. This could be a problem when important steps are to be made in healthcare, especially when it involves life-and-death decisions. Furthermore, there is also a limitation of ChatGPT fails to be a fact checker. Again, that will hamper the trust that patients and doctors have while using this powerful device.

Although these algorithm are not perfect, they are already much better than the existing chatbot. There are already 100 million users in 2 months since it is launched. One can foresee that, just as 20 years ago, we started to see patients coming in to the clinic with their Google search results about their symptoms and preferred medication; in the coming years, patient may come in with much more information and recommendations from ChatGPT. Doctors and allied health workers need to be prepared to this disruptive.

Successful implementation of AI in medicine

It is not technology which is a limiting factor rather it is the implementation of technology that is our hurdle. The introduction of Intelligent Health System into the traditional healthcare system involves a lot of changes in people's thinking, training, and workflow. It also involves a lot of patient's perspectives: trust, dignity, and autonomy. While the implementation of AI in the healthcare enterprise is complex, the ultimate goal is simple and unchanged: to help humans to do a better job of keeping people healthy and offering better care even when a cure is not possible. As technology mimics human functions such as speech, vision, and cognition come into play, they become more like human and take over activities of lower values or repetitive task. This is an inevitable change that needs to be addressed by policymakers and leaders in healthcare. If this is done right, it should free healthcare workers from mundane duties such as performing unnecessary test and investigation, reading monotonous radiological or histological images, and filling in legal forms and documents. It will allow more time to do what they want to do, i.e., listening to patients' need, answering their questions, and providing them with options and choices.

"A great place to start in pursuit of AI innovation is to shift the focus from whether AI will replace humans in doing a job, to how it will replace repetitive jobs unnecessary to be done by human or make humans better at performing specific tasks." In the book Machine, Platform and Crowd: Harnessing our Digital Future, written by MIT economist Andrew MacAfee and Erik Brynjolfsson, they modify the famous Kennedy quote "So we should ask not 'What will technology do to us?' but rather 'What do we want to do with technology.'"

Obstacles to overcome for implementation of AI in medicine

In order for AI to be able to successfully implemented and utilized in healthcare, a number of social, ethical, and legal issues should be addressed. Eight challenges that need to be addressed and overcome are

1. Transparency: Transparency refers to intelligible and understandable algorithms, processes, and outcome that can be interpreted by human intelligence. Transparency requires sufficient information of the algorithm, training data used, relevant training parameters, and results to be documented before the deployment of an AI technology.

2. Autonomy: Autonomy refers to the right of patients and healthcare providers to use or refuse to use decision-making or recommendation processes offered by AI tools. The principle of autonomy requires that machine intelligence will not undermine human autonomy.
3. Privacy: Human data, especially health data in this context, should be protected and not be used without the consent of individuals who own that data. The terms of the use of data are defined by the data protection laws of countries and regions to safeguard the use of data.
4. Equity: Equitable access and use, irrespective of age, gender, ethnicity, income, or ability to use technology to ensure inclusiveness and benefit to all.
5. Integration: For AI to be used in clinical management, AI tools should be integrated into the workflow of doctors and healthcare provider during their decision-making process.
6. Evidence: Evidence should be generated from clinical trials and outcome-based studies to ensure AI tools are validated and can provide benefits to patients and/or the healthcare system as a whole
7. Governance: Governance refers to laws and regulation to govern the use of data, a regulatory system for the approval of AI tools, and continuous monitoring of the safety and benefit of AI tools.
8. Liability: When errors and mistakes occur in the application of AI in clinical management, laws and regulations should be in place to apportion responsibility and compensation for any loss.

Engaging the clinicians and healthcare providers

The primary goal of AI is to empower clinicians, allowing them to operate at a higher level of expertise and provide better care to patients. Therefore, the key to success of AI is to gain the trust of clinicians. In order to do so, there are several steps that need to be attained.

First, a solid proof that AI-assisted diagnosis and AI-guided therapy will translate into better clinical outcome of patients. This is not just to show that certain arbitrary scoring system measuring health and recovery are improved, but robust data of improved survival, reduced hospital stay and/or clinic attendance, minimized surgery, or ICU care will be convincing evidence that machine and human working together is a better strategy.

Second, AI will displace or enhance certain types of repetitive work in specialty such as radiology and histopathology. Instead of replacing radiologists and histopathologists, there needs to be evidence to show that AI will free these experts from repetitive task and focus on higher-value activities.

Third, machine learning, deep learning, convoluted neural network, and other AI technology are Greek to most clinicians. They need to be trained and acquire at least some basic concept of these technology. They need to understand that AI recognizes pattern that may not be known to biologists or conventional clinical experts because they involve multiple layers or levels of sophisticated interpretation, amalgamation, and algorithm development. The black box in AI medicine needs to be at least partially opened to biological mechanisms and understanding so that clinicians can convey the message to their patients that the decision is based on solid ground.

Many clinicians are feeling insecure about the reliability of machine-directed management as well as the future of their job. Many are concerned also about how it will impact their autonomy, professional stature, and income. When AI is introduced into the process of caring and curing, it is meant to assist, not to replace, the healthcare provider in making decisions and carrying out treatment procedures. This underpins the importance of integrating doctors and machines working together. Patients value the doctor-patient relationship, and this should not be jeopardized because of the use of machines in the care process. In order to introduce AI successfully into daily practice of medicine, early engagement of clinicians, or even at the stage of medical education, would be needed.

Successful implementation of AI in medicine

AI will have an important role to play in the healthcare offerings, and the impact will be seen in the next 5–10 years. Given the rapid advances in AI for imaging analysis, most radiology and pathology images will be examined at some point by a machine. Speech and text recognition is already employed for tasks like patient communication and capture of clinical notes, and their usage will increase. The recent launch of ChatGPT opens a new page for diagnosis and clinical management, and its capability is astounding.

The greatest challenge to AI in these healthcare domains is not whether the technologies will be capable enough to be useful, but rather ensuring their adoption in daily clinical practice. For widespread adoption to take place, AI systems must be approved by regulators, integrated with EHR systems, standardized to a sufficient degree that similar products work in a similar fashion, taught to clinicians, paid for by public or private payer organizations, and updated over time in the field. These challenges will ultimately be overcome, but they will take much longer to do so than it will take for the technologies themselves to mature. In order to be successfully used in daily clinical practice, AI must follow the ethical principles that the outcome of AI-assisted practice and decision-making should (1) considering patients' best interests, (2) respecting the autonomy and values of patients, and (3) serving justice for all. Building on these foundations, the five pillars of a solid establishment should include (1) transparency of development and use of AI tools is provided, (2) equity of healthcare and accessibility to all social groups is guaranteed, (3) solid evidence of clinical benefit to the patients is produced, (4) privacy of patient's health and personal data is safeguarded, and (5) governance structure and regulatory framework are established.

It is undesirable that AI systems will replace human clinicians on a large scale but rather will augment their efforts to care for patients. Over time, human clinicians may move toward tasks and responsibilities that draw on uniquely human skills such as empathy, persuasion of patients, and human-machine integration (Fig. 1.2).

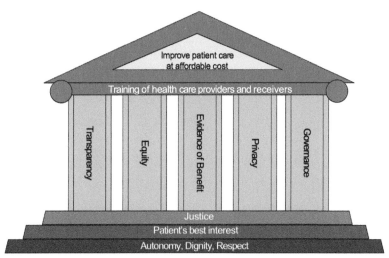

Fig. 1.2 Successful implementation of AI in medicine.

References

[1] Topol EJ. The creative destruction of medicine: How the digital revolution will create better health care. 1st pbk. ed. New York: Basic Books; 2013.

[2] Ardila D, Kiraly AP, Bharadwaj S, Choi B, Reicher JJ, Peng L, et al. End-to-end lung cancer screening with three-dimensional deep learning on low-dose chest computed tomography. Nat Med 2019;25:954–61. https://doi.org/10.1038/s41591-019-0447-x.

[3] Rodriguez-Ruiz A, Lång K, Gubern-Merida A, Broeders M, Gennaro G, Clauser P, et al. Stand-alone artificial intelligence for breast cancer detection in mammography: comparison with 101 radiologists. J Natl Cancer Inst 2019;111:916–22. https://doi.org/10.1093/jnci/djy222.

[4] Shmatko A, Ghaffari Laleh N, Gerstung M, Kather JN. Artificial intelligence in histopathology: enhancing cancer research and clinical oncology. Nat Can 2022;3:1026–38. https://doi.org/10.1038/s43018-022-00436-4.

[5] Mitchell P, Liew G, Gopinath B, Wong TY. Age-related macular degeneration. Lancet 2018;392:1147–59. https://doi.org/10.1016/S0140-6736(18)31550-2.

[6] Esteva A, Kuprel B, Novoa RA, Ko J, Swetter SM, Blau HM, et al. Dermatologist-level classification of skin cancer with deep neural networks. Nature 2017;542:115–8. https://doi.org/10.1038/nature21056.

[7] Kröner PT, Engels MM, Glicksberg BS, Johnson KW, Mzaik O, van Hooft JE, et al. Artificial intelligence in gastroenterology: a state-of-the-art review. World J Gastroenterol 2021;27:6794–824. https://doi.org/10.3748/wjg.v27.i40.6794.

[8] Johnson KB, Wei W-Q, Weeraratne D, Frisse ME, Misulis K, Rhee K, et al. Precision medicine, AI, and the future of personalized health care. Clin Transl Sci 2021;14:86–93. https://doi.org/10.1111/cts.12884.

[9] Caminiti C, Meschi T, Braglia L, Diodati F, Iezzi E, Marcomini B, et al. Reducing unnecessary hospital days to improve quality of care through physician accountability: a cluster randomised trial. BMC Health Serv Res 2013;13:14. https://doi.org/10.1186/1472-6963-13-14.

CHAPTER 2

Data access, data bias, data equity

Dennis L. Shung[a] and Loren Laine[a,b]
[a]Yale School of Medicine, New Haven, CT, United States
[b]Veterans Administration Connecticut Healthcare System, West Haven, CT, United States

Introduction

In medicine, data and the algorithms used to analyze data provide views into human health, with the explicit goal to alleviate suffering. However, all data are not created equal. The ideal for data is to reflect reality or what is actually happening.

Algorithms are well-defined procedures for carrying out computational tasks and typically are designed to analyze specific types of data. Machine learning interventions, where machine learning algorithms are applied to medical applications, typically aim to provide a diagnosis or a prediction for relevant patient outcomes to guide clinical decision-making. In order to ensure that these interventions help, and do not harm, patients, the underlying data used to train and validate these tools must be carefully considered.

Data access, data bias, and data equity are three areas that directly impact the use of and trust in machine learning interventions. They are particularly relevant if and when algorithms err, and accountability is necessary to evaluate how the algorithmic output failed to accurately reflect reality.

In order to better conceptualize how these concepts come into play, we present the hypothetical scenario of a patient, Timothy, a 56-year-old gentleman who presents to the emergency room with shortness of breath and found to be positive for COVID-19. Upon evaluation in the emergency department, an algorithm used to assess the severity of his disease finds him to be at low risk, and he is discharged from the hospital for outpatient care and quarantine. However, his clinical condition deteriorates, resulting in hospitalization and requiring mechanical ventilation in the intensive care unit.

For this particular case, data access is necessary to assess for the absence of relevant data, errors in data measurement, or errors in data storage at the time of the algorithmic prediction. An evaluation of data bias would assess whether specific heuristics learned by the algorithm contributed to the

Artificial Intelligence in Medicine
https://doi.org/10.1016/B978-0-323-95068-8.00002-9

erroneous prediction. Finally, data equity would specifically evaluate any disparate performance that may be due to Timothy belonging to an underserved or underrepresented social group.

Definition of terms

"*Data access*" refers to the unrestricted ability of a designated party to view, test, and change the raw data used to train machine learning algorithms; entities with access may be accountable for instances when algorithms err.

Healthcare data are sensitive and subject to special restrictions to ensure security and privacy. On a high level, the governance structure for "big data" or "artificial intelligence" is important in understanding the parameters of data access across societies. For clinical providers using machine learning interventions, data access refers to the entity with the permission to examine the underlying raw data used to train the algorithm. As such, if a decrease in performance is noted with errors, the entity with access to the underlying data is responsible for identifying the root cause of the error. When algorithms err, the solution starts with an overall governance strategy (who is required or encouraged to have data access). These strategies, when dealing with sensitive health data, must consider a framework for data security and privacy to manage the data access itself. Finally, data access for researchers is crucial for continuous discovery and anticipatory evaluation of potential issues with the underlying data.

"*Data bias*" identifies areas in the underlying data that may lead to differences in performance due to systemic flaws that may not reflect the best ethical or clinical practice. This can alert users of the algorithm to settings or populations where machine learning intervention recommendations should be interpreted with caution.

"*Data equity*" involves the concept of algorithmic fairness on a community and societal level, which could affect provider trust in machine learning interventions.

When an algorithm demonstrates poor performance that potentially contributes to adverse events for patients, it is essential to identify the parties responsible for accessing the underlying data to begin the investigation process so that further harm can be mitigated. Apart from identifying the specific entities responsible for this, such as the data scientists who developed the algorithm at the academic center or industry firm or governmental regulatory body, the investigation must include an understanding of both data bias

Fig. 2.1 Data access by entities with relevant expertise effectively mediates between uncovering issues with data equity and data bias and trust/accountability.

and effect of data equity. This process will help providers interpret and trust algorithmic output when applied across different settings in clinical care.

The majority of data used for clinical and translational machine learning tools can be captured in the electronic health record as part of routine clinical care. However, additional sources of data, such as patient-generated data through wearables, mobile applications, and ambient sensors, are increasingly prevalent in the healthcare data ecosystem. While the specific relevance of access, bias, and equity may differ across these different data modalities, there are also shared general principles when using health-related data (Fig. 2.1).

Data access

Governance of data

Governance encompasses how data access is managed across the entire life cycle of machine learning-based tools. Industry self-governance differs from government conceptions, which may vary by level (national government and supra-governmental organizations) and region [1–3]. For clinical providers in different regions of the world, it may be instructive to consider that issues with performance requiring access to the underlying data may lie with industry firms in the United States, the market regulatory authority in Europe, and the central government agency in China.

There is limited information about the consensus to data access across the different settings. A representative publication about industry self-governance does not mention data access specifically but assents to the best practice of data transparency and reporting. For US governmental FDA guidance, no explicit reference is given to data access, but voluntary collaboration is suggested when piloting machine learning-based interventions to assess real-world performance [4].

A paper written by Chinese authors proposed a framework for big data governance for health information networks that does not explicitly address data access, but in a series of guidelines suggests centralization and integration on a national scale in conjunction with the healthcare industry [1]. Similarly, the European Union Proposal for Regulation Laying Down Harmonized Rules on Artificial Intelligence clearly articulates the data access for training, validation, and testing dataset to be given to market surveillance authorities [5].

Security and privacy

Data access should be checked and balanced by the need for security and privacy, particularly in the case of sensitive healthcare data. The Health Insurance Portability and Accountability Act (HIPAA) applies specific requirements to all projects that are considered human subjects research, and the EU Data Privacy Regulation (GDPR) is a broader set of regulatory guidance that has specific protections for personal data [6]. HIPAA-related data protections include informed consent (or an explicit waiver with justification) and deidentification with safe harbor (removal of 18 specified personal identifiers) and expert determination [7]. In light of these requirements and concerns about the risk of reidentification, measures to maintain data security and personal privacy are an essential part of granting data access.

One framework has been proposed by the National Institute of Standards and Technology (NIST) at the US Department of Commerce to manage the risk of healthcare data: Identify, Protect, Detect, Respond, and Recover [8]. To ensure that only persons or entities with adequate training and credentials have access to sensitive data, it is important to identify data, personnel, devices, systems, and facilities pertaining to health-related data. By cataloging, maintaining an inventory, and mapping data flows as well as the roles and responsibilities for the workforce, healthcare organizations can steward access and manage the risk of a data breach or unauthorized access.

Equally important is the role of protecting access using identity management, authentication, and access control enabled by security protocols and contingency planning. When a breach of privacy occurs, whether through inadequate deidentification or an adversarial attack, detecting the incident through continuous monitoring and planned, coordinated responses with mitigation strategies are key in containing the damage.

Finally, the recovery of data and planning to incorporate lessons learned into the existing infrastructure can help safeguard the system from future incidents. For special populations, such as veterans receiving care through the United States Department of Veterans Affairs, the challenge of using sensitive data to drive innovations that may improve care delivery may have additional barriers to data access.

Data for research use

While data access is generally relevant to practitioners as users of machine learning interventions, researchers also seek data access for algorithmic development or independent validation. Not directly related to the operational mission of obtaining data access to mitigate risk in cases of errors, data access for researchers can helps anticipate problems with the underlying datasets that anticipate systemic biases in the data. In particular, research that seeks to identify disparities in access may be helpful in anticipating algorithmic errors due to underrepresentation. While access to institution-specific datasets or publicly available deidentified datasets is possible, researchers may seek more representative datasets that are not limited to a specific center or health system.

Two datasets that have well-defined security protocols and rich longitudinal data, the All of Us Research Program (an initiative of the National Institutes of Health) and the UK Biobank, are high-profile efforts to build a diverse health database representative of modern US and UK society, respectively [9,10]. When thinking through data access, the mission of Fig. 2.2

Fig. 2.2 Extract page of All of Us Research Program, National Institutes of Health (https://allofus.nih.gov).

Fig. 2.3 Extract page of UK Biobank (https://www.ukbiobank.ac.uk).

encompasses not only university researchers but also citizen scientists and administrators interested in quality improvement initiatives.

Fig. 2.3 furthers the access to specifically include those in low- and middle-income countries who do not traditionally have the capability of acquiring such large-scale and in-depth data. These initiatives provide a valuable opportunity for motivated researchers to identify systemic sources of bias and characterize underlying data trends that may be useful when troubleshooting a particular episode of algorithmic error.

Data bias

A critical piece of evaluating algorithmic performance is the identification of data bias, which should consider both data available for initial model development and the ongoing data being analyzed by the algorithm in practice. An analogy for this approach is that of the physician-patient relationship, where after the initial appointment there is an ongoing relationship that periodically reevaluates the first diagnoses and revises them as more information is obtained.

Bias is not just present in the data used for the model; the absence of specific data elements due to distrust or lack of resources, changes in practice patterns or billing, and inappropriate inclusion of race in recommended decision tools could influence the model performance [11]. Finally, the increasingly common data sources of ambient sensors and mobile applications can lead to bias due to the exclusion of entire sectors of society from the digital front door due to incomplete rollout, poor uptake, and the diffusion of responsibility between the private and public sectors.

A key question that should be asked throughout the life cycle is the following: do choices made about the data that are measured or captured during the life cycle worsen or perpetuate existing health inequalities? [12]. As the Oslerian ideal of equanimity, or mental equilibrium, motivates providers to consider clinical findings above and beyond the temptation to fall into specific cognitive biases, this question should motivate physicians, researchers, and data analysts to go from focusing on getting as much data as possible to critically considering the characteristics and deficiencies of the data being used.

Electronic health record data

Data collected in the electronic health record or as part of routine clinical care are generated in the context of clinical medicine, where the patient-physician relationship should be considered as a framework for understanding bias. In this framework, the patient may not tell the physician all the relevant information, may seek care with other physicians, and may not have the means to undergo the recommended testing. The physician may have a unique diagnostic or treatment style and may perceive patient reports through the lens of their experience and perception.

Missing data

Missing data can be a source of bias, but are only addressed or accounted for in 54% of predictive algorithmic studies using electronic health record data [13]. Missingness could be due to the absence of provider entry into the health record, an error of omission, or secondary to lack of access to diagnostic tests or procedures [12,14]. Lack of access has been documented in surgical care and endoscopic access, particularly impacting patients identifying as Black [15,16].

Fragmentation of care without systemic interoperability can contribute to data missingness, particularly if they are linked to other societal factors such as low socioeconomical status, psychosocial issues, or immigration status [17,18]. In particular, demographic and socioeconomic data are often incomplete, with an estimated one-third of commercial insurance plans reporting complete or partially complete data on race [18].

Variation in practice patterns

Since practice may change across different providers and health systems, differential care patterns can lead to misclassification and measurement error [12]. This may be seen across teaching and nonteaching settings or across urban and rural areas and correspond with uninsured patients or patients on Medicaid, which has been well documented in the emergency department setting [19].

Provider bias

Patient language is usually communicated and filtered through providers, who then enter the information into the electronic health record. As such, the physician may selectively record information according to their clinical experience and unfortunately, sometimes their biases.

In Psychiatry, where patient-reported data are primarily filtered through providers, this has been seen as potentially problematic in the stage between the expression of data by the patient and the interpretation of data by the provider; data captured by the provider and then analyzed by a machine learning algorithm to predict an outcome can magnify the bias [20]. For example, if providers prescribe higher doses of psychiatric medications to patients with minority race backgrounds despite having similar reported symptoms, an algorithm could then be biased to predict higher doses of necessary medication by race. Another example is that because women are more likely to receive personality disorder diagnoses compared to men, when women present with the same symptoms of trauma, an algorithm may perpetuate this diagnostic pattern as predicted diagnoses [21–23].

Explicit racial correction in clinical tools

The historical practice of medicine has incorporated race explicitly in ways that purposely bias calculations of clinically significant measures. Corrections in definitions of organ performance, such as the glomerular filtration rate, heart failure mortality risk, and pulmonary function tests, may be based on race. For example, Vyas et al. recently identified 13 clinical algorithms in which race was explicitly used to modulate risk and determine courses of treatment [24]. When used in clinical practice, these algorithms may lead to disparities in access to care for patients identifying with a specific primary race. For example, the heart failure risk score recommended by the American Heart Association Guidelines increases the risk for "nonblack" patients when deciding referral to a cardiology specialist when hospitalized [25]. This guideline-recommended score, if used in clinical care, could result in Black patients being deemed to be at lower risk despite having the same other characteristics of a nonblack patient with subsequent lack of access to specialty care.

Wearables and ambient sensors

Data captured from electronic health records are primarily filtered through provider entry and health system priorities, whereas wearable and app-captured data reflect direct patient input and monitoring of physical environments by ambient sensors. However, the issue of bias can also impact these data sources that exclude patients from the "digital front door." This can be along similar lines of resource access due to socioeconomic status and optimization of these devices for specific populations or as in the case of the recent effort by the United Kingdom's National Health Service to roll out a mobile app due to botched rollouts and poor uptake [26]. In digital behavioral-change interventions to increase physical activity, the devices appear to have greater effectiveness for people with higher socioeconomic status [27]. For wearables tracking health-related information (e.g., heart rhythms, oxygenation, sleep patterns), the challenge of the sensor technology for maintaining accuracy across skin pigmentation tones as well as the absence of diverse representation in validation studies may lead to differential effectiveness by race and ethnicity [28].

For ambient sensors in health settings, the measurement or capture of data on multiple participants in different contexts can have the potential for reinforcing biases seen in one context that are problematic in another. Behavior between participants in a certain context may have a specific interpretation that may bias the interpretation of algorithms trained on this setting when transferred to another setting [29]. For example, a sensor-based algorithm trained to identify provider activity in an intensive care unit where congregation of multiple people indicates increased clinical concern for patient deterioration may not be directly applicable to a psychiatric ward, where congregation of multiple members of the care team may be part of a therapeutic group session.

Data equity

Fairness for algorithms has been defined by three pillars: transparency, impartiality, and inclusion [30]. Practical categories such as geographic region, socioeconomic strata, gender, and race/ethnicity can be used as a starting point for thinking about overall fairness. Explicitly addressing these categories may provide a basis for provider trust in the applicability of these tools in practice.

Transparency

Transparency includes interpretability, explainability, and accountability. Interpretability can be defined using performance metrics such as accuracy, sensitivity, and specificity [31]. These metrics can give a general idea about how well the algorithm can identify patterns, which is the necessary first step before considering how to integrate the machine learning intervention into clinical care.

Explainability displays data elements that are used for the algorithmic output and also includes the specific context from which the data are generated to give providers additional information about the relevance of the algorithmic output. By understanding the setting from which training data are generated, providers using the tool can consider whether specific factors may lead to nonequitable algorithmic outputs. For example, Black patients presenting with acute upper gastrointestinal bleeding to emergency departments in the United States have lower odds of receiving an upper endoscopy; if the code for previous endoscopy is used as an input variable to a machine learning intervention to predict the need for urgent endoscopic evaluation for patients with acute upper gastrointestinal bleeding, the bias against getting an endoscopy for Black patients may be amplified [32].

Finally, accountability refers to the question of responsibility in the event of an adverse outcome. A clear assignment of responsibility must be defined to mitigate the risk to providers using these tools. When thinking through accountability in healthcare, it may be useful to think of the machine learning intervention in terms of a consultation to access additional information or expertise a provider cannot otherwise access. There should be a clear designation of responsibility in the event of an adverse event in which the machine learning intervention generates the wrong prediction. In order to understand the role of the machine learning intervention in contributing to the adverse event, providers should know who to consult as part of the root cause analysis.

Impartiality

Impartiality includes provenance (the origins and characteristics of the data) and implementation. The starting point is clinical relevance in the specific setting of use to ensure that the patient population in the intended setting is represented by the data used to train the machine learning intervention [33]. Then, during implementation, the anticipated harms should be considered to evaluate if there is a disproportionate impact on specific populations. This process should include evaluation of discriminatory practices that arise or are exacerbated by the integration of the machine learning intervention. For example, in a modeling analysis, machine learning-based predictive tools for medical appointment scheduling may amplify the higher no-show probability for Black versus non-Black patients and lead to wait times up to 30% longer for Black patients due to recommendations for overbooked appointments [34].

Inclusion

Inclusion encompasses data completeness and utilization of traditionally excluded data sources, such as patient-reported or community-reported data. Data completeness should consider geographic distribution and representativeness, which is currently skewed in machine learning applications for clinical medicine to disproportionately used cohorts from California, Massachusetts, and New York [35]. Race-based differences may be more relevant in societies with historical policies of purposely disadvantaging specific racial groups, though other categories may better reflect the specific historical disparities of each society. For example, across Latin America, the concept of *mestizaje*, or fusion of indigenous, European, and African heritage due to the mixture of colonialism, intermarriage, and slavery, leads to a complex set of identities that lead to social inequalities [36].

Algorithmic stewardship as a framework for mitigation strategy against bias and inequity

The essential part of strategies to identify and address issues with algorithmic bias is the human in the loop. While data access is essential for any mitigation strategy, the study and monitoring of the real-world effectiveness of algorithmic machine learning interventions are arguably more important in identifying areas of adverse effect possibly attributed to bias or inequity.

Algorithmic stewardship encompasses the necessary practices to mitigate bias and test for inequity [37]. Algorithm vigilance, which refers to methods for evaluation, monitoring, understanding, and preventing adverse effects of healthcare algorithms, is a useful framework for the development and deployment of informatics-based methods for debiasing algorithms [38]. The main parts of stewardship include the creation and maintenance of algorithm inventories, an auditing process prior to deployment, and periodic review by a group with oversight responsibility. While this can be thought of at a health-system level, it is likely that governmental or supra-governmental support may be necessary due to the limited expertise available at a health-system level to understand and evaluate algorithmic performance with clinical relevance.

In the case of the algorithmic error for our patient Timothy, the monitoring system should identify the error during routine evaluation of algorithmic performance as measured by the proportion of inaccurate predictions. Those with designated data access can then evaluate the performance to evaluate the role of bias (specific data elements that were not captured) or equity (consistently depressed performance among patients like Timothy).

The need for an interdisciplinary approach, including sociological and ethics expertise, is critical since algorithms are typically designed to maximize performance and efficiency and not necessarily to reflect human values. Both Eastern and Western traditions have emphasized the human-centered character of medical practice; Confucius is quoted to have said, "medicine is a humane art," and the Hippocratic Oath exhorts physicians to "first, do no harm" [39]. The cost of imposing human values may result in decreased efficiency; in medicine, however, all decisions should be filtered through the human-centered values of the medical profession [12]. The challenge of ensuring that this is enforced as a special part of artificial intelligence ethics is compounded by the increasing dominance of industry in artificial intelligence, as documented by the Artificial Intelligence Index Report 2022 [40].

The socio-technical challenge is particularly acute in healthcare, where trust is paramount, and the stakes are high. The involvement of physicians and healthcare providers as algorithmic stewards who actively participate in the design, evaluation, and implementation of machine learning interventions is critical to ensure that efficiency does not justify losing sight of humanistic values. By hewing to the human-centered vision, algorithmic stewards can guide decisions regarding data access, bias, and equity that are consistent with the spirit of medical care.

References

[1] Li Q, Lan L, Zeng N, You L, Yin J, Zhou X, et al. A framework for big data governance to advance RHINs: a case study of China. IEEE Access 2019;7:50330–8. https://doi.org/10.1109/ACCESS.2019.2910838.
[2] Roski J, Maier EJ, Vigilante K, Kane EA, Matheny ME. Enhancing trust in AI through industry self-governance. J Am Med Inform Assoc 2021;28:1582–90. https://doi.org/10.1093/jamia/ocab065.

[3] Vokinger KN, Gasser U. Regulating AI in medicine in the United States and Europe. Nat Mach Intell 2021;3:738–9. https://doi.org/10.1038/s42256-021-00386-z.

[4] The U.S. Food and Drug Administration. Artificial intelligence and machine learning in software as a medical device; 2021 [Accessed 30 March 2022].

[5] European Commission. A European approach to artificial intelligence; 2022 [Accessed 30 March 2022].

[6] Chico V. The impact of the general data protection regulation on health research. Br Med Bull 2018;128:109–18. https://doi.org/10.1093/bmb/ldy038.

[7] Committee on Health Research and the Privacy of Health Information: The HIPAA Privacy Rule, Board on Health Sciences Policy, Board on Health Care Services, Institute of Medicine. Beyond the HIPAA privacy rule: Enhancing privacy, improving health through research. Washington, DC: National Academies Press; 2009.

[8] The National Institute of Standards and Technology. NIST cybersecurity framework; 2018 [Accessed 30 March 2022].

[9] All of us research program; 2022 [Accessed 30 March 2022].

[10] UK Biobank 2022. (Accessed 30 March 2022).

[11] Thomasian NM, Eickhoff C, Adashi EY. Advancing health equity with artificial intelligence. J Public Health Policy 2021;42:602–11. https://doi.org/10.1057/s41271-021-00319-5.

[12] Gianfrancesco MA, Tamang S, Yazdany J, Schmajuk G. Potential biases in machine learning algorithms using electronic health record data. JAMA Intern Med 2018;178:1544. https://doi.org/10.1001/jamainternmed.2018.3763.

[13] Goldstein BA, Navar AM, Pencina MJ, Ioannidis JPA. Opportunities and challenges in developing risk prediction models with electronic health records data: a systematic review. J Am Med Inform Assoc 2017;24:198–208. https://doi.org/10.1093/jamia/ocw042.

[14] Johnson-Mann CN, Loftus TJ, Bihorac A. Equity and artificial intelligence in surgical care. JAMA Surg 2021;156:509. https://doi.org/10.1001/jamasurg.2020.7208.

[15] Abougergi MS, Avila P, Saltzman JR. Impact of insurance status and race on outcomes in nonvariceal upper gastrointestinal hemorrhage: a nationwide analysis. J Clin Gastroenterol 2019;53:e12–8. https://doi.org/10.1097/MCG.0000000000000909.

[16] Haider AH, Scott VK, Rehman KA, Velopulos C, Bentley JM, Cornwell EE, et al. Racial disparities in surgical care and outcomes in the United States: a comprehensive review of patient, provider, and systemic factors. J Am Coll Surg 2013;216:482–492e12. https://doi.org/10.1016/j.jamcollsurg.2012.11.014.

[17] Arpey NC, Gaglioti AH, Rosenbaum ME. How socioeconomic status affects patient perceptions of health care: a qualitative study. J Prim Care Community Health 2017;8:169–75. https://doi.org/10.1177/2150131917697439.

[18] Ng JH, Ye F, Ward LM, Haffer SCC, Scholle SH. Data on race, ethnicity, and language largely incomplete for managed care plan members. Health Aff 2017;36:548–52. https://doi.org/10.1377/hlthaff.2016.1044.

[19] Greenwood-Ericksen MB, Kocher K. Trends in emergency department use by rural and urban populations in the United States. JAMA Netw Open 2019;2, e191919. https://doi.org/10.1001/jamanetworkopen.2019.1919.

[20] Straw I, Callison-Burch C, Danforth CM. Artificial intelligence in mental health and the biases of language based models. PLoS ONE 2020;15, e0240376. https://doi.org/10.1371/journal.pone.0240376.

[21] Becker D, Lamb S. Sex bias in the diagnosis of borderline personality disorder and post-traumatic stress disorder. Prof Psychol Res Pract 1994;25:55–61. https://doi.org/10.1037/0735-7028.25.1.55.

[22] Jane JS, Oltmanns TF, South SC, Turkheimer E. Gender bias in diagnostic criteria for personality disorders: an item response theory analysis. J Abnorm Psychol 2007;116:166–75. https://doi.org/10.1037/0021-843X.116.1.166.

[23] Snowden LR. Bias in mental health assessment and intervention: theory and evidence. Am J Public Health 2003;93:239–43. https://doi.org/10.2105/AJPH.93.2.239.

[24] Vyas DA, Eisenstein LG, Jones DS, Malina D. Hidden in plain sight—reconsidering the use of race correction in clinical algorithms. N Engl J Med 2020;383:874–82. https://doi.org/10.1056/NEJMms2004740.

[25] Peterson PN, Rumsfeld JS, Liang L, Albert NM, Hernandez AF, Peterson ED, et al. A validated risk score for in-hospital mortality in patients with heart failure from the American Heart Association get with the guidelines program. Circ Cardiovasc Qual Outcomes 2010;3:25–32. https://doi.org/10.1161/CIRCOUTCOMES.109.854877.

[26] Best J. The NHS App: opening the NHS's new digital "front door" to the private sector. BMJ 2019; l6210. https://doi.org/10.1136/bmj.l6210.

[27] Western MJ, Armstrong MEG, Islam I, Morgan K, Jones UF, Kelson MJ. The effectiveness of digital interventions for increasing physical activity in individuals of low socioeconomic status: a systematic review and meta-analysis. Int J Behav Nutr Phys Act 2021;18:148. https://doi.org/10.1186/s12966-021-01218-4.

[28] Colvonen PJ, DeYoung PN, Bosompra N-OA, Owens RL. Limiting racial disparities and bias for wearable devices in health science research. Sleep 2020;43, zsaa159. https://doi.org/10.1093/sleep/zsaa159.

[29] Martinez-Martin N, Luo Z, Kaushal A, Adeli E, Haque A, Kelly SS, et al. Ethical issues in using ambient intelligence in health-care settings. Lancet Digit Health 2021;3: e115–23. https://doi.org/10.1016/S2589-7500(20)30275-2.

[30] Sikstrom L, Maslej MM, Hui K, Findlay Z, Buchman DZ, Hill SL. Conceptualising fairness: three pillars for medical algorithms and health equity. BMJ Health Care Inform 2022;29, e100459. https://doi.org/10.1136/bmjhci-2021-100459.

[31] Rajkomar A, Hardt M, Howell MD, Corrado G, Chin MH. Ensuring fairness in machine learning to advance health equity. Ann Intern Med 2018;169:866. https://doi.org/10.7326/M18-1990.

[32] Rodriguez NJ, Zheng N, Mezzacappa C, Canavan M, Laine L, Shung D. Disparities in access to endoscopy for patients with upper gastrointestinal bleeding presenting to emergency departments. Gastroenterology 2022; S001650852201157X. https://doi.org/10.1053/j.gastro.2022.10.001.

[33] Oala L, Murchison AG, Balachandran P, Choudhary S, Fehr J, Leite AW, et al. Machine learning for health: algorithm auditing & quality control. J Med Syst 2021;45:105. https://doi.org/10.1007/s10916-021-01783-y.

[34] Samorani M, Harris SL, Blount LG, Lu H, Santoro MA. Overbooked and overlooked: machine learning and racial bias in medical appointment scheduling. Manuf Serv Oper Manag 2019. https://doi.org/10.13140/RG.2.2.11208.06404.

[35] Kaushal A, Altman R, Langlotz C. Geographic distribution of US cohorts used to train deep learning algorithms. JAMA 2020;324:1212. https://doi.org/10.1001/jama.2020.12067.

[36] Rodríguez ME. How the mixed-race mestizo myth warped science in Latin America. Nature 2021;600:374–8. https://doi.org/10.1038/d41586-021-03622-z.

[37] Eaneff S, Obermeyer Z, Butte AJ. The case for algorithmic stewardship for artificial intelligence and machine learning technologies. JAMA 2020;324:1397. https://doi.org/10.1001/jama.2020.9371.

[38] Embi PJ. Algorithmovigilance—advancing methods to analyze and monitor artificial intelligence–driven health care for effectiveness and equity. JAMA Netw Open 2021;4, e214622. https://doi.org/10.1001/jamanetworkopen.2021.4622.

[39] Zhang D, Cheng Z. Medicine is a humane art. The basic principles of professional ethics in Chinese medicine. Hast Cent Rep 2000;30:S8–12.

[40] Stanford University. AI Index Report 2022; 2022 [Accessed 30 March 2022].

CHAPTER 3

Respect for persons

Tamra Lysaght[a], Vicki Xafis[b], and Cameron Stewart[c]
[a]Faculty of Medicine and Health, Sydney School of Public Health, Sydney Health Ethics, University of Sydney, Sydney, NSW, Australia
[b]University of Melbourne, Melbourne, VIC, Australia
[c]Sydney Law School, University of Sydney, Sydney, NSW, Australia

Introduction

As artificial intelligence (AI) technologies are increasingly entering health-care settings, the need for values-based decision-making processes is becoming more salient. AI systems are currently being developed and implemented in a variety of domains that include, but are not limited to, clinical diagnosis, treatment and predictive modeling, operational management, and patient monitoring with sensory systems and the internet of things. Numerous ethical issues surrounding the use of AI systems have been articulated [1–3], and a substantial number of ethical principles and guidance documents have been published for the development and implementation of AI [4]. These documents include ones specifically targeted at healthcare such as the *Ethics & Governance of Artificial Intelligence for Health* from the World Health Organization (WHO) [5], *Ethics Guidelines for Trustworthy AI* (European Commission) [6], the *WMA Statement on Augmented Intelligence in Health Care* from the American Medical Association [7], OECD Health Working Paper 128: *Laying the foundations for artificial intelligence in health* [8], and the UNESCO *Recommendation on the ethics of artificial intelligence* [9]. Specialist colleges around the world, for example the UK Royal College of Physicians [10] and the Royal Australian and New Zealand College of Radiologists, have also started developing position statements and guidance for the use of AI in health [11].

These documents all recognize a number of important ethical principles. Transparency, accountability and responsibility, justice and fairness, nonmaleficence, and privacy are some of the ethical principles that have been proposed along with solidarity, sustainability, human dignity, and respect for persons [4]. However, tensions arising between conflicting interests and values in the use of AI can be expected to vary across domains and according to the healthcare setting in which the systems are applied. For example, the

Artificial Intelligence in Medicine
https://doi.org/10.1016/B978-0-323-95068-8.00003-0

27

use of AI systems to support decision-making in emergency healthcare settings, where time is of the essence, may raise different concerns when used
for chronic disease management, surgical interventions, or embryo selection
for fertility treatments. This variation demands more than reference to
abstract principles as concrete steps on how to apply them to the specific circumstance are required.

To meet this demand, we draw on a step-by-step deliberative process for
ethical decision-making that was developed by a working group convened
by the *Science, Health and Policy-relevant Ethics in Singapore* (SHAPES) Initiative, at the National University of Singapore in 2019. This working group,
of which the authors of this chapter were members, developed the *Ethics
Framework for Big Data in Health and Research* [12] (hereafter the Framework)
and applied it to six domains where the use and application of big data raises
ethical concerns, including AI in healthcare [13]. The Framework is a tool
for deliberating on issues and bringing to the fore, relevant values which
ought to guide or "frame" decision-making [12], as distinguished from the-
orizing or modeling [14]. It sets out a collection of substantive and procedural values that can be identified as being in conflict or tension, and then
weighed up in a six-step deliberative process. It also articulates three overarching ethical concerns that apply across domains and warrant special consideration: vulnerability of data contributors and their communities, the
social license or legitimacy of big data systems operating without explicit
consent from individuals, and, perhaps most importantly, respect for
persons.

This chapter expands on the application of the SHAPES *Ethics Framework
for Big Data in Health and Research* to the domain of AI in healthcare with
special consideration of the overarching issue of *respect for persons*. Respect
for persons is the moral attitude of placing the interests and values of other
human beings at the center of decision-making. Respect for persons is demonstrated by how we act upon, or omit to act upon, the interests and values
of others during our interactions with individuals, groups, or institutions.

The chapter first describes the six-step deliberative process and summa-
rizes the relevant substantive and procedural values drawn from the ethics
framework for AI in healthcare. We will then discuss how respect for persons applies to the development and use of AI generally, linking the issue
with the concept of social license and the legitimacy of embedding AI systems in healthcare contexts.

Finally, we will employ the six-step deliberative process outlined by
Lysaght et al. as it relates to the case of AI-assisted provider decision support

systems (PDSS) for managing resources in intensive care units (ICUs) [13]. We will then conclude with some reflections on the Framework, with particular reference to the value of respect for persons.

A values-based deliberative balancing approach to ethical decision-making

An ethical decision-making framework is a conceptual tool that helps its users think through an ethically complex issue, or set of issues, in a structured and systematic manner. Use of such frameworks does not obviate the need to ensure inclusion of all relevant stakeholders, from technical experts (e.g., designers and developers), to health experts (e.g., healthcare professionals and healthcare organizations), to regulators and patients. A framework cannot "provide" the answer on its own and cannot guarantee well-considered decisions in the absence of personal and general decision-making wisdom [15,16] or the moral courage for certain decisions. Although our Framework identifies an (inexhaustive) set of substantive and procedural values as most relevant to the use of big data in numerous contexts, including AI in clinical practice, no one value is given priority at the outset [14,17], either implicitly or explicitly. To do so would consistently force decisions in a particular direction [14,17] and would therefore undermine the whole decision-making process. By the end of the process, it should only be the relevant values that are used to make decisions.

Process of identifying, considering, and resolving ethical issues

Before the process of decision-making commences, it is critical to ensure the range of expertise required is available. Community members who have a stake may also be included to ensure that decisions are in line with community expectations.

Step 1—Identifying and clearly articulating the problem or ethical issue
Clearly articulating the issue(s) will assist in identifying and clarifying some of the surrounding problems, complexities, and ethical concerns that will require consideration.

Step 2—Identify the relevant values
The specific issue/problem and the context within which it arises will help identify the relevant values. In addition, examining the three overarching considerations (respect for persons, social license, and vulnerability and

Substantive values are those considerations that should be realised through the outcome of a decision.   ➡️   Harm minimisation | Integrity |Justice | Liberty/Autonomy |Privacy | Proportionality | Public benefit | Solidarity |Stewardship

Procedural values are the values that guide the process of deliberation and decision making itself.   ➡️   Accountability | Consistency | Engagement | Reasonableness | Reflexivity | Transparency Trustworthiness

Fig. 3.1 Substantive and procedural values [12].

power) in the specific context will also help to identify some of the values we need to consider in each case. The Framework [12] itself provides a potentially relevant list of values which will simplify the process for users (Fig. 3.1).

Step 3—Identifying potential actions

There is likely to be an iterative process in steps 1, 2, and 3 before all the relevant issues and values are identified. That is, when decision-makers begin to consider possible courses of action, it is likely that additional values will become obvious and will lead them to continue to consider the specific solution, amend it, or even discard it. It is at this point that Framework users will also be guided by constraints placed upon decision-making, for example, as a result of legislative or regulatory requirements.

Step 4—Weighing up relevant ethical merit of options

Uncertainty and incomplete information may sometimes render this process challenging but decision-makers must be able to acknowledge and tolerate some level of uncertainty so as not to be frozen into inaction. Weighing up the relevant ethical merit of the options is the most challenging task because when values are in tension with each other, decision-makers will need to judge which values should take priority over others. No framework can include general guidance on how to prioritize conflicting values because it is impossible to articulate a single process for balancing these outside of a specific context.

Judging and justifying the best course of action cannot be supported by simply listing the values that would be satisfied or violated by a specific course of action. In each application of the Framework, decision-makers have to develop a convincing case for privileging some values while compromising others. Because it will never be possible to privilege all values equally, the Framework highlights the importance of always articulating mechanisms to minimize the potential harms associated with any chosen course of action. This process, in itself, may help with deciding which value(s) should carry greater weight.

Step 5—Selecting the option with strongest ethical weight and reflecting on potential influences

In step 4, decision-makers should have begun to formulate robust justifications for the various choices that they are making. In step 5, they settle on the course of action and justify the decision(s) in detail to demonstrate that they are morally coherent. Robust justifications for the rejection of alternative courses of action and trade-offs will do part of this work and will provide greater decision-making transparency. Throughout the decision-making process, but particularly in this step, it is critical to reflect on potential influencing factors biasing the process at an individual or group level.

Step 6—Communicating the decision transparently

The final step requires the transparent communication of the decision. It must also be borne in mind that the decision reached may need to be re-considered at some future point depending on the circumstances. For example, rapid technological developments may require some adjustments (Fig. 3.2).

Application of the framework to AI in healthcare

In the application of the Framework to AI in healthcare, Lysaght et al. identified five substantive and procedural values as being the most relevant:

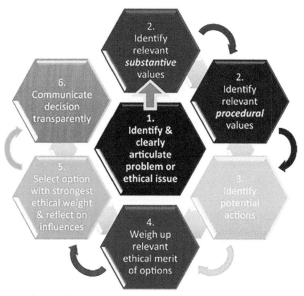

Fig. 3.2 Process of identifying, considering, and resolving ethical issues [12].

justice, public benefit, (professional) integrity, transparency, and accountability [13]. However, the selection of values was only intended to provide a demonstration of how the different values might apply so was not exhaustive in its treatment. Here, we identify additional values that may have relevance in other healthcare contexts where AI systems are being developed and implemented. This list is primarily drawn from Xafis et al. [12] with some adaptation. Additionally, the list also broadly correlates with ethico-legal principles previously identified as relevant in healthcare settings [18].

Relevant substantive values for AI in the healthcare context include:

Beneficence: Beneficence is concerned with improving the life of individuals and groups. This value is centrally important as the use of AI in healthcare is targeted at improving the ability of health professionals to diagnose, prognosticate, and treat human illness [3].

Harm minimization: This value is based on the ethical principle of non-maleficence [3,6] and it seeks to reduce "the possibility of real or perceived harms (physical, economic, psychological, emotional, or reputational) to persons" [12] as well as their social groups and/or communities. AI systems ought to be designed and implemented in ways that minimizes harms for individuals, particularly for vulnerable and marginalized communities who may well gain the least benefits from AI as they may be under-represented in the training datasets and have generally fewer encounters with the health systems AI is designed to support.

Integrity: This refers to the professional "values and commitments" that health professionals can be expected to follow in their practice. It is important for health professionals to maintain integrity when interacting with and incorporating AI-supported systems into their practice, and doing so without compromising their primary moral and legal responsibilities to patients [13]. This can apply to anyone who in positions of trust with fiduciary duties to persons AI-supported decision-making may affect.

Liberty/autonomy: Liberty can be defined as a state of freedom from the coercion of physical, legal, or social pressures. Autonomy refers to the ability a person or group to make decisions for themselves. Patients must be able to make decisions, communicate those decisions, and have those decisions respected. When offering AI-assisted treatment to patients, health professionals need to allow patients to apply their own values to questions of consent and participation. Health professionals also need to be sensitive to how patients frame their own desires and values with their communities and it should not be assumed that there will be a uniform community acceptance and understanding of AI-use in healthcare.

Privacy/Confidentiality: In this work, privacy refers to the right to have some control over access to information about oneself and the right not to be re-identified and adversely impacted by the synthesis of stored data. Confidentiality is a related concept that requires sensitive information to only be used in ways that have been agreed to. Both values are relevant to AI as AI is empowered by the use of health data, especially combinations of datasets that may not have been traditionally linked. Traditionally, health data are protected by principles of use-purpose limitation, data scarcity, and use minimization. However, AI requires access and use of health data that is open and able to be linked across numerous datasets. AI usage of health information should always be based on consent, or legally justified alternatives (such as approvals from data access controllers and human research ethics committees). Such processes require the interests of patients to be considered and protected against privacy-based harms such as reputational damage, ridicule, or hatred.

Justice: The value of justice is concerned with treating people fairly, consistently, and with respect. Justice is also concerned with the fair distribution of benefits and burdens and with equity. One of the major ethical concerns with AI is the risk of discriminatory harms arising from AI systems replicating and potentially exacerbating social biases and prejudices that may be built into and/or subsequently introduced into training datasets [13]. Justice as fairness requires decision-makers to examine how the benefits and the burdens of AI systems might be distributed equitably across the population.

Public Benefit: This value relates to the principle of beneficence but is more broadly construed as "the overall good that society as a whole receives" [12]. It includes the effects on "well-being, distribution, societal cohesion, human rights, and other sources of value" [12]. Such benefits may include fewer medical errors and more efficient healthcare systems. These benefits should be balanced carefully against possible burdens to both individual patients and marginalized groups that AI may further disadvantage [13].

Procedural values that are relevant to AI in healthcare include:

Reflexivity: This procedural value refers to the need to reflect on and respond to the limitations and uncertainties embedded in knowledge, information, evidence, and data. It includes being alert to competing and conflicting personal, professional, and organizational interests and to the management of associated biases. Reflexive institutions should recognize the risks of biases becoming entrenched in the AI algorithms and build in processes to identify and remove them from their systems as soon as possible. Attending to this value may help to mitigate against potential injustices brought about by the replication of biases in AI systems.

Accountability: This value relates to the principles of responsibility and liability. It refers to the "ability to scrutinize judgements, decisions and actions, and for decision-makers to be held responsible for their consequences" [12]. Developers of AI may be held responsible for ensuring that the systems are programmed to be explainable and minimize social biases and prejudices, although medical professionals should arguably remain liable for clinical decisions that affect patients under their care [13].

Transparency: This value broadly relates to ensuring that "decision-making, processes, and actions" are open to public scrutiny and helps to "demonstrate respect for persons and contributes to trustworthiness" [12] It relates to concerns about the increasing complexity of AI algorithms and challenges for decision-makers in understanding how the system makes recommendations, especially when they run contrary to their own professional judgments [13]. AI algorithms that are explainable, or able to be scrutinized, may help promote transparency and trustworthiness [6].

Trustworthiness: This value is the "property of being worthy of trust" that applies to individuals, organizations, governments, and institutions, as well as to data, evidence, and systems [12]. Trust may also be placed in processes that are transparent and truthful, reliable and consistent, and dependable [12] AI-assisted health systems will be highly reliant on the trust of health professionals and patients for their successful implementation and adoption.

Engagement: This value is crucial for enacting trustworthiness and maintaining a social license. At an individual level, engagement pertains to obtaining meaningful consent from patients, especially during the development and implementation phases [3]. At the population level, it requires "meaningful involvement of stakeholders in the design and conduct" of AI systems [12].

Respect as an overarching consideration

What is respect for persons? We have elsewhere defined *respect* as relating to "one's moral attitude toward others and the actions toward others that result from and exemplify this attitude" [12]. Here we elaborate on this core concept of "respect." In doing so, we do not attempt to unravel the complexities of numerous theories expounding the conditions that are necessary/sufficient to ground the kinds of moral status that give rise to respect [18,19]. Respect always has an object, i.e., the recipient of respect [20], and in this work the object is *persons*. Hence, we accept the traditional account of moral status based on human properties [18] but stress that we support the proposition that "all beings with moral status, even though they may have less than full status, still have *some* moral status" [18] and should, therefore, be afforded respect.

Treatment of respect in the literature is *other-regarding*, that is, it has as a starting point acknowledging, paying attention to, or heeding the object (here person) of respect [21–29]. However, we depart from this standard view in a small but significant way. The essence of respect, in our view, is a consideration of one's own moral worth as being equal to that of others'; that is, not viewing our own moral worth, values, views, judgments, projects, etc., as superior to others'. In line with Anderson [21], as expressed by Bejan, our conception of this moral attitude relates to a *relation of equivalence* [30] of moral status of persons even if there are numerous differences exhibited. It is this fundamental attitude of nonsuperiority that underlies and enables us to recognise and value other people's intrinsic worth.

Our conception of respect is therefore diametrically opposed to the manner in which Lysaught appears to view respect: "To treat another with respect, … is to put them above and ahead of ourselves" [29]. In our understanding of this core essence of respect, the attitude or stance we hold does not relate to an evaluation of the object of our respect, either positive or negative, but rather, "as something whose significance is independent of us" [20] but nevertheless of relational significance. Attitudinal coherence demands that we recognize the universal nature of respect; in other words, that like objects warranting respect be treated in like ways [20], all things being equal.

Analyses of respect [31] offer a variety of classifications in an attempt to better understand the nature of respect. For example, Darwall identifies *recognition respect* ("appropriate consideration or recognition to some feature of its object … take seriously and weigh appropriately the fact that they are persons" p. 38) and *appraisal respect* ("a positive appraisal of a person, or his qualities" p. 39) [23] in the contemporary sense of "admire" [32]. Our work does not deal with the latter. Hudson, on the other hand, offers four narrow classifications of respect, centered on the basis for which respect is shown: *evaluative respect* (similar to Darwall's *appraisal respect*); *obstacle respect*;

directive respect; and *institutional respect* [27]. In some contexts, this classification might be useful even though three of the classifications Hudson proposes appear to be reducible to what Darwall considers *recognition respect*, a point also highlighted by Dillon [20].

We do not view behavior or actions toward others as an element, component, or dimension of respect [31] or as a form of respect itself [20]. Rather, we view behavior arising from an attitudinal stance of respect as the operationalization of respect. We contend that behaviors and responses arising from our attitudinal stance may or may not always adequately convey respect because how we signal respect is grounded in social and cultural practices that may not be shared. To provide a simple example, in many cultures, looking directly at the speaker with whom you are engaging is a sign of respect, as it demonstrates that you are attentive to their contributions and are duly noting them. In some cultures, however, extended direct gaze could be viewed as disrespectful [33].

In many bioethics discussions, the notion of respect has been subsumed under the concept of *respect for autonomy* [18,34] but this considerably narrows the scope [28] because respect for autonomy relates only to those who have the capacity to make decisions in line with their views, wishes, and values. Our broader view of respect incorporates the narrower notion of respect for autonomy and thus accounts for those whose decision-making capacity may be reduced, fluctuating, or nonexistent, often the case in the healthcare context; it also amply accommodates vulnerable persons or groups, whose autonomy may be affected. To respect in this broad sense is to consider seriously what aspects of life are valuable to people (or were valuable to people who can longer express these themselves), what promotes their projects in life, and the socially derived values and practices that bring meaning to their lives (on the understanding that the above do not interfere with others' interests). This broader view of respect is of particular importance when considering AI applications in healthcare, as considerations of respect for autonomy alone [35] could obscure other important considerations.

Positive and negative obligations arise from what we view as the core concept of respect. Positive obligations require the taking of action to recognize and protect people's intrinsic worth. Negative obligations require us to refrain from taking actions that fail to take account of people's intrinsic worth or that interfere with people's ability to live life in the way they consider best. In this sense, respect is restrictive of behaviors toward the object of our respect [23]. Obligations, however, are requirements to which only moral agents are bound. While thought-provoking discussions around technology and agency are now being considered [35], in this work we view the designers, developers, and implementors of AI technology (and policy-makers) as directly or indirectly accountable for actions and harms arising from algorithms and machine learning [5,6,9].

The substantive and procedural values considered in this work all promote respect and autonomy in varying ways. Hence, we regard respect as an overarching, foundational value.

Having justified our conceptualization of respect and related values, we will now apply them to the case study of AI-assisted Provider Decisions Support Systems (PDSS) for managing resources in intensive care units (ICUs). PDSS is a term we use to broadly capture the types of decision-making systems that AI can inform or *augment* beyond the conventional clinical decision support systems (CDSS) that were the focus of the Lysaght et al case discussion [13]. CDSS have been used to assist physicians in forming diagnoses and predicting clinical outcomes for decades and the literature has previously examined the ethics of practitioners using these tools [36]. Traditional CDSS are hard-coded software programs that are based on static databases. They are often patented, which facilitates public scrutiny of their methodology. AI-driven systems, by contrast, are deeply complex and evolving. Their immense computational power can take into account much wider, varied and linked sources of information—from electronic medical records, genome sequences, clinical trial databases and surveillance systems, to wearable and sensing devices. These systems can also introduce feedback in real time and *learn* to support not only clinical decisions but also operational management processes and policy formation [37]. Our case discussion will apply the Framework to the management of AI-assisted PDSS in ICUs.

Case study: Applying the Framework to AI-assisted PDSS

A software developer approaches the head of an intensive care unit (ICU) in a large university hospital to build an AI-assisted PDSS that can predict medical futility. The Application ("App") will not only help tertiary care physicians predict the outcomes of patients admitted to the ICU with much higher accuracy (98%) but will help hospital administrators allocate resources according to the volume of patients that the system predicts will be admitted to the unit at any one time. One of the issues the head of this ICU considers important is whether an App that can essentially predict the death of patients in critical and end of life (EOL) care should be introduced into the clinical setting and, if so, how much the clinicians should rely on the App's recommendations. She is also concerned about liability and the impact the system may have on existing consent-taking procedures in the ICU.

Identifying the problems and issues

There are numerous ethical issues raised by the case study. Increasing accuracy of predicting a patient's future course in ICU may improve care at a number of levels. At the patient and family level, there is potential for improvements to be made in prognostication so that patients are more likely to receive treatments that will match their best interests. Families may also be able to be better prepared for poorer outcomes, such as death or disability. At the health practitioner level, doctors and allied health professionals will be able to give more accurate information to patients and their families and will be able to provide more targeted treatments or palliative care. The App has the potential to help clinicians rationalize the weighing of clinical factors with their professional expertise and experience. It may give them more confidence and support in accurately predicting outcomes in ICU. At the unit or hospital level, there is potential for better resource management, more targeted use of resources for better patient outcomes and less waste.

However, there are areas of concern. Firstly, there is a potential for the App to displace the clinical judgments of the treatment team if the App is not used appropriately. The treatment team may be tempted to rely too heavily on the App for decision-making and clinical decision-making skills may atrophy. There may be a risk that the treatment team "outsources" the ethical burden of decision-making to the App.

Equally problematic would be a circumstance where the App's predictions unexpectedly conflict with the clinical assessments made by the treatment team. This may be less of an issue if the App's logic algorithm is transparent and understood by the treatment team. However, if the App's decision-making processes are not easily comprehended and understood by the treatment team it will be very difficult for the team to reassess their decision and reflect on why the App's recommendations are different from theirs.

At a deeper level, the App developers' claim to predict "medical futility" almost certainly misapplies this concept. The term itself is contested because it is ill-defined and because agreement on a set of criteria that adequately assess a futility threshold has not been reached [38,39]. Medical futility is not reducible to a prognosis on which a decision to pursue or not pursue a treatment hinges [39]. Rather, it involves an ethical judgment about whether a course of treatment is "worth" pursuing from a range of perspectives, including, but not limited to, clinical outcomes (which themselves can be conceptualized in a variety of ways, as aptly demonstrated by Wilkinson and Savulescu [38]. Hence, this judgment should involve a very careful assessment of the patient's interests and wishes, which would most likely necessitate careful discussion with the patient and their family about the treatment goals [39]. The App, therefore, presents the risk that other ethically relevant considerations relating to medical futility and its assessment could be ignored, hidden or displaced.

Resource allocation decisions are a potential source of liability but they are usually defensible when it can be shown that the treatment team used established protocols that were accepted by peers as competent professional practice [39]. If it was shown that patients were denied treatment because the treatment team had relied on the App, and not on established standards of care, claims for negligent harm may well be brought by patients and their families. This is less likely to be an issue if the App was shown to have been used as part of a futility assessment process that was built on competent professional practice and supported by peer review [40].

Identifying the relevant substantive ethical values

Looking first at the substantive values, *beneficence* is relevant if it can be demonstrated that the App improves decision-making for patients, health care professionals, and administrators. *Harm minimization* is also relevant as the App will help to avoid errors and waste. But it is also important to recognize the App's potential for harm, especially given that ICU patients are particularly vulnerable and that training datasets may not be wide and deep enough to account for significant differences in cultural and ethnic groups [41]. The App's design may be able to be adapted to be conscious of these variables and include them in its learning systems.

As discussed above, *integrity* is also a pressing concern, as the App should be understood as enhancing professional judgment, rather than supplanting it. Improving the quality of ICU care will improve trust in the profession and the underpinning notion that decisions will be made on the best scientific evidence. The care that doctors provide to patients and the respect that these professionals demonstrate toward the ICU patient and their family by treating the patient with dignity, contributes to their professional integrity. The App could also provide a structure for conversations about the best interests of patients, including advice on what questions to ask and on how to communicate difficult news. While the family's values and wishes should be considered and respected, professional integrity requires that doctors do not yield to family demands to implement highly invasive treatments that will not benefit the patient and that may cause unnecessary pain or the prolongation of the dying process. These problems may be countered by carefully tracking how the treatment team employs the App to see whether it is being used to enhance or supplant decision-making [13].

Liberty and *autonomy* are also engaged and are particularly important values for consent processes in the ICU. It is vital that patients and their family members will be informed about how the App will be used to aid decision-making. It is also important to communicate how the patient's health information will be employed by the App, not only for the immediate decision of what treatments may be available to the patient, but for how the information will feed into the App's learning processes in the future. This issue directly involves the values of *privacy* and *confidentiality*, because of the way it engages concerns over control, access, and use of heath information.

Justice is a more abstract value in this scenario, but one can see it becoming relevant when the App starts to make recommendations about futility. As discussed above, such use would be inappropriate. While the App may be able to provide highly accurate prognostication, it is unable to provide an ethical judgment on the patient's best interests, a judgment that involves a wider range of considerations than simply identifying a particular treatment option and the threshold for it to be regarded futile. It is important that the treatment team not "outsource" the ethical burden of decision-making to the App, as discussed above, as this would offend the notion that patients need to be treated fairly, consistently, and with respect.

Finally, *public benefit* is also a highly relevant consideration. At the *micro* level, the App has the potential to improve the quality of care, speeding up the time for decision-making, and giving more confidence to the treatment team. At the *meso* level, the App may improve resource usage in the ICU unit, which may help other patients in the ICU. At the *macro* level, the use of the App may increase efficiencies in resource allocation at the hospital and in the local health service. Improvements in intensive care may also translate to greater levels of trust in the health system, greater engagement with patient communities, and better public understanding of how ICU decision-making occurs [13].

Identifying the relevant procedural ethical issues

Reflexivity is an important value as the App needs to be designed to proactively monitor how differences in treatment options and outcomes occur across different socio-economic, racial, and ethno-cultural groups [41].

Transparency and *accountability* are also critically important. The developers of the App need to be able to explain how the App uses AI to generate a prognosis, especially if the App appears to be providing answers that may be at odds with the treatment team's assessments. Patients and their families also need to be told about the App and how it is used in decision-making. These discussions should clearly state why the App is being used and how it is not being used.

Accountability also requires that health professionals maintain their professional responsibility, even when they have used an App in their decision-making. As discussed above, the treatment team must avoid outsourcing the ethical burden of decision-making to the App. By making sure that the App is a reliable tool used by the treatment team in their decision-making processes (rather than as a replacement for those processes) the risk of liability is reduced. If it is shown that the App actually does improve the prognostication of the treatment team then it cannot be argued that it has harmed the interests of the patients. Morever, using the App may reduce the risk of liability as patient outcomes are improved.

These concerns also engage the value of *trustworthiness* as transparent and accountable processes in App use with allow for patients and families to trust that their interests are being properly assessed and that decisions are based on sound considerations. The presence of trust may further reduce the risk of patients (and their family members) seeking legal redress for wrongdoing.

Engagement is also crucial for the App to be able to grow in accuracy and predictive power. Engagement with the patients (and their families when necessary) is important for creating meaningful consent, strengthening trustworthiness, and maintaining the social license for the App use [3]. If the App later shows that there are groups that have poorer outcomes in ICU, engagement with those groups will be necessary to find the causes of the poorer outcomes. This will not only increase the kinds of benefits that the App can produce but it will help to improve harm minimization and broader issues of justice.

Returning to respect for persons

This discussion of substantive and procedural values highlights the centrality of respect for persons to the case study. The decision to employ the App in the ICU has clear benefits for individual patients, the medical team, and the hospital. However, the core focus of decision-making must be the patient's best interests, whether that manifests in active treatment or managed withdrawal and palliative care. The focus on best interests is a manifestation of respect for persons. Building systems of transparency, accountability, and engagement is also respectful of persons as it values their concerns and inputs in the continued improvement of the predictive power of the App. Concern for the wider issues of improving resource allocation is also a form of respect for persons as it improves the health system's capacity to care for future patients.

Conclusion

This chapter applied a values-based deliberative framework to the use of AI-assisted decision-making tools in healthcare. Artificial intelligence (AI) systems are being developed to assist the decision-making of healthcare professionals, administrators, and researchers in wide-ranging settings from diagnosis to treatment, patient selection, and policymaking. While guiding ethical norms and principles for AI use are well-articulated, their application to particular settings will be context-specific and depend on variables that may change how competing values come into tension and are balanced to arrive at morally justifiable decision-making. We applied a six-step deliberative process that aims to balance relevant substantive and procedural values to support ethical decision-making with AI in healthcare settings. We employed a case study of AI-assisted PDSS in intensive care resource management to consider how the framework might usefully address issues of substantive and procedural ethical values. We argued that underlying all of these values is a core concern for respect for persons. Any framework for consideration of AI systems in healthcare should build upon that core value.

References

[1] Beil M, Proft I, van Heerden D, Sviri S, van Heerden PV. Ethical considerations about artificial intelligence for prognostication in intensive care. Intensive Care Med Exp 2019;7:70. https://doi.org/10.1186/s40635-019-0286-6.
[2] Cath C. Governing artificial intelligence: ethical, legal and technical opportunities and challenges. Philos Trans R Soc A Math Phys Eng Sci 2018;376:20180080. https://doi.org/10.1098/rsta.2018.0080.
[3] Stewart C, Wong SKY, Sung JJY. Mapping ethico-legal principles for the use of artificial intelligence in gastroenterology. J Gastroenterol Hepatol 2021;36:1143–8. https://doi.org/10.1111/jgh.15521.
[4] Jobin A, Ienca M, Vayena E. The global landscape of AI ethics guidelines. Nat Mach Intell 2019;1:389–99. https://doi.org/10.1038/s42256-019-0088-2.
[5] World Health Organisation. Ethics and governance of artificial intelligence for health: WHO guidance; 2021.

[6] European Commission. Directorate general for communications networks C and technology, high level expert group on artificial intelligence. Ethics guidelines for trustworthy AI. LU: Publications Office; 2019.

[7] World Medical Association. WMA statement on augmented intelligence in medical care., 2019, https://www.wma.net/policies-post/wma-statement-on-augmented-intelligence-in-medical-care/. [Accessed 28 April 2023].

[8] Hashiguchi TCO, Slawomirski L, Oderkirk J. Laying the foundations for artificial intelligence in health; 2021.

[9] United Nations Educational Scientific and Cultural Organization. Recommendation on the ethics of artificial intelligence n.d.

[10] Royal College of Physicians (UK). Position statement on artificial intelligence (AI) in health, https://www.rcplondon.ac.uk/projects/outputs/artificial-intelligence-ai-health. [Accessed 21 March 2022].

[11] Royal Australian and New Zealand College of Radiologists. Standards of practice for clinical radiology., 2022, https://www.ranzcr.com/college/document-library/ranzcr-standards-of-practice-for-diagnostic-and-interventional-radiology, 11.2. [Accessed 21 March 2022].

[12] Xafis V, Schaefer GO, Labude MK, Brassington I, Ballantyne A, Lim HY, et al. An ethics framework for big data in health and research. Asian Bioeth Rev 2019;11:227–54. https://doi.org/10.1007/s41649-019-00099-x.

[13] Lysaght T, Lim HY, Xafis V, Ngiam KY. AI-assisted decision-making in healthcare: the application of an ethics framework for big data in health and research. Asian Bioeth Rev 2019;11:299–314. https://doi.org/10.1007/s41649-019-00096-0.

[14] Dawson A, Peckham S, Hann A. Theory and practice in public health ethics: a complex relationship. In: Public health ethics and practice. Policy Press; 2009. p. 191–210.

[15] Staudinger UM, Glück J. Psychological wisdom research: commonalities and differences in a growing field. Annu Rev Psychol 2011;62:215–41. https://doi.org/10.1146/annurev.psych.121208.131659.

[16] Staudinger UM, Ferrari M, Weststrate NM. The need to distinguish personal from general wisdom: a short history and empirical evidence. In: The scientific study of personal wisdom. Dordrecht: Springer Netherlands; 2013. p. 3–19.

[17] Grill K, Dawson A. Ethical frameworks in public health decision-making: defending a value-based and pluralist approach. Health Care Anal 2017;25:291–307. https://doi.org/10.1007/s10728-015-0299-6.

[18] Beauchamp TL, Childress JF. Principles of biomedical ethics. 7th ed. New York: Oxford University Press; 2013.

[19] Agnieszka J, Tannenbaum J, Zalta EN, Nodelman U. The grounds of moral status. In: The Stanford encyclopedia of philosophy; 2021.

[20] Dillon RS, Zalta EN, Nodelman U. Respect. In: The Stanford Encyclopedia of Philosophy; 2022.

[21] Anderson ES. What is the point of equality? Ethics 1999;109:287–337. https://doi.org/10.1086/233897.

[22] Beach MC, Duggan PS, Cassel CK, Geller G. What does 'respect' mean? Exploring the moral obligation of health professionals to respect patients. J Gen Intern Med 2007;22:692–5. https://doi.org/10.1007/s11606-006-0054-7.

[23] Darwall SL. Two kinds of respect. Ethics 1977;88:36–49. https://doi.org/10.1086/292054.

[24] Dickert NW, Kass NE. Understanding respect: learning from patients. J Med Ethics 2009;35:419–23. https://doi.org/10.1136/jme.2008.027235.

[25] Dillon RS. Respect for persons, identity, and information technology. Ethics Inf Technol 2010;12:17–28. https://doi.org/10.1007/s10676-009-9188-8.

[26] Henry LM, Rushton C, Beach MC, Faden R. Respect and dignity: a conceptual model for patients in the intensive care unit. Narrat Inq Bioeth 2015;5:5A–14A. https://doi.org/10.1353/nib.2015.0007.

[27] Hudson SD, Department of Philosophy FSU. The nature of respect. Soc Theory Pract 1980;6:69–90. https://doi.org/10.5840/soctheorpract19806112.

[28] Kraft SA, Rothwell E, Shah SK, Duenas DM, Lewis H, Muessig K, et al. Demonstrating 'respect for persons' in clinical research: findings from qualitative interviews with diverse genomics research participants. J Med Ethics 2021;47:e8. https://doi.org/10.1136/medethics-2020-106440.

[29] Lysaught M. Respect: or, how respect for persons became respect for autonomy. J Med Philos 2004;29:665–80. https://doi.org/10.1080/03605310490883028.

[30] Bejan TM. What was the point of equality? Am J Polit Sci 2022;66:604–16. https://doi.org/10.1111/ajps.12667.

[31] Subramani S, Biller-Andorno N. Revisiting respect for persons: conceptual analysis and implications for clinical practice. Med Health Care Philos 2022;25:351–60. https://doi.org/10.1007/s11019-022-10079-y.

[32] Cambridge University Press. Cambridge Dictionary., 2022, https://dictionary.cambridge.org/dictionary/english/. [Accessed 23 April 2022].

[33] Uono S, Hietanen JK, Hadjikhani N. Eye contact perception in the west and east: a cross-cultural study. PLoS ONE 2015;10, e0118094. https://doi.org/10.1371/journal.pone.0118094.

[34] Beauchamp TL, Ashcroft RE, Dawson A, Draper H, McMillan JR. The 'four principles' approach to health care ethics. In: Principles of health care ethics. Chichester, UK: John Wiley & Sons, Ltd; 2006. p. 3–10.

[35] Laitinen A, Sahlgren O. AI systems and respect for human autonomy. Front Artif Intell 2021;4, 705164. https://doi.org/10.3389/frai.2021.705164.

[36] Goodman KW, Berner ES. Ethical and legal issues in decision support. In: Clinical decision support systems. New York, NY: Springer; 2007. p. 126–39.

[37] Liu N, Zhang Z, Wah Ho AF, Ong MEH. Artificial intelligence in emergency medicine. J Emerg Crit Care Med 2018;2:82. https://doi.org/10.21037/jeccm.2018.10.08. https://jeccm.amegroups.com/article/view/4700.

[38] Wilkinson D, Savulescu J. Knowing when to stop: futility in the ICU. Curr Opin Anaesthesiol 2011;24(2):160–5. https://doi.org/10.1097/ACO.0b013e328343c5af.

[39] Stewart C. Futility determination as a process: problems with medical sovereignty, legal issues and the strengths and weakness of the procedural approach. J Bioeth Inq 2011;8(2):155–63.

[40] Stewart C, Kim P. The standard of care test revisited: competing approaches to defining competent profession practice in Australia. J Law Med 2023;30:278–85.

[41] Lee JJ, Long AC, Curtis JR, Engelberg RA. The influence of race/ethnicity and education on family ratings of the quality of dying in the ICU. J Pain Symptom Manag 2016;51:9–16. https://doi.org/10.1016/j.jpainsymman.2015.08.008.

CHAPTER 4

Privacy and confidentiality

Stephen Kai Yi Wong
Gilt Chambers, 8/F Far East Finance Centre, Central, Hong Kong

Introduction

When a patient visits or is ferried to a clinic, hospital, or health care (med-health) institution, the first thing doctors, nursing, and health care staff (med-health professionals) will have to do is to take down personal health information (PHI) about him. Normally a form will be filled in, whether electronically or on a piece of paper. Admittedly, the purpose is to identify him for the purposes of medical examination, diagnosis and providing the appropriate treatment, medications, or health care. This set of PHI is in law known as "personal data" or "personal information." The patient is the "data subject," and the med-health institution, professional or any related third party is the "data user," "data controller," "data processor," or collectively the "data handler."

If the patient is consciously providing his PHI, he does it voluntarily and with his consent. If his Glasgow Index (a measurement of consciousness level) is low, his PHI, albeit limited, is collected and used for emergency or life-saving purposes. The privacy law says the patient should remain in control of his own PHI. The law of confidentiality says no one other than the med-health professionals should know about his PHI. The ethics say the med-health professionals should do what they believe they ought to do for this patient and in the circumstances. That said, the laws and ethics do not prohibit but regulate the collection, retention and subsequent use, disclosure, or transfer of the PHI on the part of the med-health professionals.

All med-health professionals have the legal and ethical responsibility to respect and protect PHI. Patients have the right to protection of their own PHI, as well as an expectation that it will be held in confidence. Members of the public may also have an interest in the PHI in situations involving public health or crime. Balancing properly the competing interests in PHI is ultimately conducive to building trust and providing quality services.

Artificial Intelligence in Medicine
https://doi.org/10.1016/B978-0-323-95068-8.00004-2

Personal health information (PHI)

PHI is often statutorily defined or classified as sensitive information. Despite the fine legalistic distinction between privacy personal information and confidential information (which will be used interchangeably in this chapter), examples of PHI include:

- personal biographical or personal history data;
- biometrics (biologic measurement or physical characteristics which can help to identify an individual, e.g., fingerprint, retinal scan, or facial image);
- previous contacts and locations records (especially for epidemics and pandemics);
- medical history or records (including family members' medical records);
- medical care information (examinations, diagnoses, assessments, and service reports);
- guardianship orders or relevant persons' information for the incompetent or incapacitated;
- relevant personal correspondence; and
- payment method and account.

Biometric data need a particular mention. In the context of privacy law [1], it is a "special" personal data resulting from specific technical processing relating to the physical, physiological, or behavioral characteristics of a natural person, which confirms the unique identification of that natural person, such as facial recognition or fingerprint data. The processing of this unique data is generally prohibited, subject to exemptions.

For regulatory compliance and ethical considerations, the challenges of handling PHI have been intensified by the use of electronic health records, mobile health devices, and artificial intelligence (AI) in med–health sciences and services.

Privacy

Privacy is generally recognized as a fundamental human right, a right that is qualified and not absolute in nature. What it means is that a person's information is free from unlawful interference by the others, persons, or organizations alike. A person is entitled to the control and protection of his own personal data, including his physical whereabouts, subject to qualifications (or exemptions) set out in the laws. He can, for example, set privacy restrictions on his social networking site account to limit the access to his personal information like who can see it, profile pictures, etc. Should there be a clash between a patient's qualified privacy right and other competing rights (such as public health or other public interests) or an absolute right (such as others' right to life), there may be a case for him to have to surrender, voluntarily or otherwise, his own individual privacy right.

Confidentiality

Confidentiality is about information agreed to be kept secret between the relevant parties, or to be shared with only a few people for a designated purpose unless the person to whom the information belongs permits to disclose it. That is where industry- or profession-specific laws and regulations including professional code of conduct, as well as a confidentiality contract, would come in to prevent misuse or abuse. In the case of a confidentiality agreement, the relationship between the relevant parties is a fiduciary one. A fiduciary duty is owed by the one receiving the information who acts on behalf of the other, normally a client, putting the client's interests above his, to preserve good faith and trust. Confidentiality in daily lives includes one's bank account number, ATM pin, user password of an email account. While confidentiality is protected by or enshrined in laws and court decisions, it is not, unlike privacy, generally regarded as a human or constitutional right.

Privacy and confidentiality in medical and health services

Privacy in the context of med-health services means that when a patient tells his med-health professionals, who will in turn create a file that includes information about any tests, treatment, and medication they give, will be respected and protected in the course of it being collected, stored, used or shared, and accessed. Patients may give their consent to share their PHI, e.g., when they change med-health institution and the med-health professionals there should have access to their medical history.

Confidentiality in med-health services means that the med-health professionals keep the PHI in confidence between themselves and the patient. They have a legal and ethical duty not to divulge it to anyone other than those who need to know, unless the patient consents or disclosure is required by law or is necessary for the public interest. This also means not showing anyone—again, other than those who need to know—personal notes or computer records about the patient. The doctor-patient relationship establishes an implied contract of confidentiality since the doctor is in a position to collect and use PHI.

Privacy and confidentiality are at the center of maintaining trust between patients and med-health professionals. The doctor-patient confidentiality practically manifests the core values of respecting and protecting patients' personal data privacy right in that the med-health professionals undertake and patients are assured that his PHI will not be disclosed to others without their consent. This element of trust is fundamental to the therapeutic nature of the doctor-patient relationship. It is by no means merely a matter of compliance but serves as an incentive for patients to seek care openly and confidently, especially in cases where data sensitivity or personal embarrassment is involved.

Classes of patients may also call for special attention. For example, in certain jurisdictions, children may seek treatment without the permission of their parents for certain conditions, such as treatment for pregnancy, sexually transmitted infections, mental health concerns, and substantive abuse. Patients aside, family members or visitors may approach med-health professionals for information. Strictly speaking, disclosure is restricted to authorized or relevant persons (e.g., persons named as guardians or contact persons), although authorized access to PHI, eHealth records included, is often extended to the medical teams concerned and discussions among them.

Most privacy laws regulate the protection of PHI of natural or living patients only, with the latest exception being the Personal Information Protection Law (PIPL) [2,3] implemented in China in November 2021 whereby personal information may continue to be protected subject to the deceased's wish [4]. However, the duty of confidentiality to patients remains after death. In situations where, e.g., an issue arising after a patient's death, med-health professionals may be requested to discuss or disclose PHI with family members or an intimate member out of wedlock, especially if the patient was a child or a celebrity. Ultimately, it is the instructions of the deceased that should be followed. It should also be noted that as a matter of law, the personal representative of the deceased (i.e., the executor or the administrator of his legacy depending on whether he died with or without a will) is entitled to access the relevant parts of the deceased's PHI.

Privacy crossing paths with confidentiality

It is not surprising that the issues of privacy and confidentiality protection and obligations overlap. Technical contexts aside, the core principle dictates that med-health institutions and professionals are required, by law or as a matter of ethics, to demonstrate they have policies implemented to protect the privacy and confidentiality relating to PHI, including electronic information, the procedures for computer access, and security.

Unintended disclosures may occur, e.g., when a patient's case is discussed in the lift or other public places. Similarly, copies of handouts from training courses may contain identifiable patients. However, med-health professionals might not realize someone is ears dropping a discussion that includes protected PHI, especially where patients share a room. While they do not need to change the physical set up, med-health professionals do need to ensure that not just anyone can see it. They should exercise extra caution when they handle PHI which includes both paper and electronic records and steer clear of earshot and not leaving it in plain sight (Box 4.1).

BOX 4.1 Privacy and confidentiality

- Privacy is a fundamental human right—qualified but not absolute, free from unlawful interference and an entitlement to control and protection but subject to statutory exemptions including competing public interest, e.g., public health.
- Patient's PHI should be respected and protected when collected, stored, used or shared, and accessed.
- Confidentiality is a contractual arrangement whereby information is agreed to be kept secret and shared restrictedly—a fiduciary duty owed by the recipient to preserve interparties good faith and trust but not as a human or constitutional right.
- Patient's PHI collected is also protected by the implied contract of confidentiality—not to be divulged to other parties than those who need to know unless with patient's consent or required by law or necessitated in the public interest (health).
- Doctor-patient confidentiality demonstrates trust which also serves as an incentive for patients to seek care and treatment, in particular where special attention for sensitive cases is called for.
- Unlike privacy right generally, the duty of confidentiality extends beyond a patient's life unless otherwise instructed by the patient or mandated by his will or the intestacy law.
- To avoid unintended disclosures, extra caution should be exercised when handling PHI, whether in paper or electronic form, by steering clear of earshot and plain sight.

Exemptions from privacy breaches

The general principle is that where there is a legitimate reason and pressing need for regulating, the conflict between an individual's qualified right and public interest, the latter should prevail. Balancing properly, the laws across the globe provide a legal basis for it by way of setting out the exemptions from data protection breaches, i.e., no liability will result if the requirements of the exemptions are proved to have been satisfied. Scenarios where med-health professionals may use, disclose, or transfer patients' PHI include:

- when specific *consent* has been given by the patient explicitly;
- when the health or safety of the *patient*, being incapable of giving consent, is seriously at risk of harm and the information will help, e.g., where the patient is unconscious and there is a pressing need to know how his relatives can be contacted or if he is allergic to any drugs (Emergency Exemption);
- when the information will reduce or prevent a serious threat to *public* health or safety and withholding the information will expose others' health or safety to serious risk of harms and threats, e.g., where the patient has a serious contagious illness and the public needs to be warned by using location tracking devices or applications (Public Interest Exemption);
- when the information will reveal the commission of a *crime*, e.g., where the patient intends to hurt himself or others, or suspicious wounds or bruises are found on his body during a clinical examination (Crime Exemption); and
- when the information is required by *law* or is critical in any legal claims (Rule of Law Exemption).

Data protection principles (DPP)

The entire idea about data privacy protection is that the data subjects are well informed of and consent to what the data controllers/users/processors do about their personal information. Med-health professionals must therefore manage PHI in compliance with relevant DPPs that govern the collection, retention, use or disclosure, confidentiality, security and access, complemented by practising professional ethics.

Lawfulness, fairness, transparency, and explainability

Basically, no one should do anything unlawful with personal data. In this digital age, extensive and ubiquitous collection of personal data, both online and offline, together with the unpredictability in the use, disclosure, transfer, sharing, and security breach of data, individuals may not even be aware that their data have been collected or stolen, not to mention exercising control over and objecting to unfair use of it.

Starting from the surveillance cameras installed in the med-health institutions, those responsible should consider how the data handling may affect the patients and justify any adverse impact. Med-health professionals do not deceive or mislead patients when they collect their PHI and should be open, honest, and comply with the transparency obligations of patients' right to be informed, e.g., by displaying a privacy notice in the vicinity of the cameras as a starting point.

Having clear policies of dealing with PHI can help patients trust the med-health professionals more. Patients' trust is most important for retaining confidence, securing repeat visits, and showing integrity. In law, it is a requirement that patients are made aware of how their privacy is valued by way of a Privacy Policy and Personal Information Collection Statement (PICS), in which it should be stated in clear and comprehensible terms:

- what information is collected, and who this information may be shared with;
- whether other third parties may collect or use the information, and if so state who they (or the class) are; and
- how patients can request changes and access to their information.

Purpose limitation and data minimization

Patients' data are collected for a stated purpose which should be as limited as possible, never a blanket one. Only data that are actually needed for the specified purpose should be collected. This limited purpose should be essentially the same as that of its subsequent use or disclosure, and in case of a new or different purpose, prior consent will have to be obtained unless it falls within the exemptions.

Accuracy and the right to rectification and erasure

Only accurate data are collected, more so in the case of med-health services. Appropriate processes to check the accuracy of the data collected, update, and rectify should be put in place. Where appropriate, patients may request that their PHI be erased by virtue of the enhanced Right to be Forgotten.

Retention limitation

It is trite law that an organization can only store data for as long as is necessary for the stated purpose. Med-health records may serve a medicolegal purpose and therefore it is legitimate for the records of a patient to be kept even if they will not be used for the purpose of medical consultation. While PHI may also need to be retained for a relatively long time for public interest archiving, scientific or medical research, or statistical purposes, clear and justifiable procedures to govern the removal or deletion of data should be developed. One common practice is to store PHI in a limited number of systems, depending on the data type or classification. A smaller system footprint would also reduce the chance of data security breaches.

Use and disclosure

The obvious purpose of collecting PHI is to use it. Generally, use in relation to PHI includes disclosure or transfer, e.g., transferring it to another party or team for follow-up, posting it on controlled notice boards and transmitting it through internet.

If the purpose of use is the same as or directly related to the purpose for which PHI is collected, no prescribed consent is needed. In ascertaining the original purpose of collection, regard will have to be had to:
• the explicit purposes stated in the PICS;
• the functions or activities of the med-health professionals; and
• the restrictions of use imposed by the patient (if any).

Confidentiality, integrity, and data security

PHI must be handled in such a way as to ensure appropriate confidentiality, integrity, and security (e.g., by using encryption). Security goes directly to the root of PHI protection. Specifically, it relates to the means to protect PHI privacy and assist in handling PHI in confidence. Med-health records in paper form aside, there are concerns about the confidentiality of eHealth records, given their saleable and marketing value, ease (or vulnerability) of access and transmission.

Security safeguards must be implemented to ensure PHI is not exposed to unauthorized use, disclosure, access, processing, erasure, or loss, whether intentional or accidental. In recent years, ransomware attacks have become common cyber security threats. Where PHI is attached by a malware, a data breach will occur because access to the data is lost and permanent data loss can also result if there are no backups.

That said, even the most robust security practices can fail—some breaches can be caused by individuals with approved access. The risk of insider breach threats can be mitigated by ensuring that datasets are accessible only to those who need them and that no one has access to all available data.

One of the highest indicators of data subjects' trust is the speed of reporting and data handlers' response when data security breaches occur. Indeed, most regulations require reporting, to regulators and affected data subjects alike, within, e.g., 72h of the announcement of a data breach incident. To act quickly when breaches occur, med-health professionals will want to pressure-test their crisis-response processes in advance. Nearly, all data security breaches involve whether appropriate technical and organizational measures [5] are designed and implemented effectively.

Technical measures mean anything from requiring med-health professionals to use multifactor authentication on accounts where PHI is stored to using its own or contracting with cloud services providers that use end-to-end encryption.

Organizational measures are things like staff training, adding a data privacy policy (including data security policy) to the staff handbook, or limiting access to PHI to only those who need it, having particular regard to assessing the risk of inadvertent human error, including the integrity, prudence, and competence of persons having access.

Access and rectification

Patients have the right to request and receive a copy of their PHI, and other supplementary information in their medical record under the control or in the custody of med-health professionals or institution. This is commonly referred to as a data access request (DAR). Complying with a DAR calls for the understanding of what is to be delivered and how it can be refused or delivered in the correct format, normally within a prescribed period of time.

Accountability

Compliance with the DPPs aside, med-health professionals are also required to demonstrate they are compliant. The legal principle of accountability requires them to account for the risks arising from their handling of PHI—whether they are running a simple register of patients' contact details or operating a sophisticated Artificial Intelligence (AI) system in the operation theater. That is why the requirement of accountability has been introduced in many legislative frameworks to demand evidence of steps having been taken to

- adopt a "Data Protection by Design and Default" approach, i.e., adopting appropriate data protection measures before and throughout the entire data handling operations;
- carry out Data Protection Impact Assessments in a timely manner;
- designate data protection responsibilities to a professional data protection officer;
- maintain detailed documentation of the collection, retention, and use of data;
- train and implement technical and organizational security measures;
- have Data Processing Agreements in place with third parties; and
- record and, where necessary, report on data breaches.

Setting up good accountability practices is a key part of securing patients' trust and staying in line with their legal obligations. That is also why a comprehensive external Privacy Policy should be put in place, as well as internal guidance that promotes the protection and care of PHI.

Consent

Consent in the context of PHI means offering the patient a real choice and control. Like the right to be informed, the choice of giving or withdrawing consent is a core value of privacy protection. Genuine consent should be able to put the patient back on the driver seat, build trust and engagement, and enhance the institution's reputation.

A patient's consent should be given by an affirmative act establishing a freely given, specific, well informed, and unambiguous indication of his agreement to the handling of his PHI by, e.g., a written statement electronically or orally. This could include ticking a box when visiting a med-health institution's Internet website or other conduct which unequivocally indicates his acceptance of the ways his PHI will be handled. Silence, preticked boxes or inactivity do not therefore constitute consent.

Adult patients are assumed to be competent to give consent to the use of their PHI, unless there is specific reason to doubt it. When the patient is a minor, med-health professionals must consider whether he understands fully the implications of a decision regarding proposed collection, use or disclosure of his PHI.

It is common practice that PHI needs to be shared within the med-health teams. The legal basis is that the purpose is directly related to the purpose for which the PHI is collected at the outset, not that the consent is implied or can be inferred. The best practice is to de-identify the patients. Consent is presumed not to be freely given if it does not allow separate consent to be given to or withdrawn from new or different personal data handling purposes. Practically, when the patient is subsequently surprised by what is subsequently alleged to have been consented to, there was no genuine consent.

In light of the technology development, med-health professionals should obtain patients' consent for the use of electronic means and/or devices for med-health care (e.g., sending digital photographs) and for communicating PHI (e.g., the use of email). To obtain informed consent, patients should be explained, preferably in writing, about the related benefits and risks.

As data governance stewards, med-health professionals play an important role in educating patients about possible consensual and nonconsensual use and disclosure that may be made of their PHI, including its secondary use for, e.g., epidemiological studies, research, education, and quality assurance.

Enhanced privacy rights

Enhanced or new privacy rights which aim to give data subjects extended control over their own personal data have been introduced in the laws of some jurisdictions, notably Europe and China. Those relevant to med-health services include:

Right to restrict and object to processing

Patients may request the restriction or suppression of the processing of their PHI, mainly because the accuracy of the data needs to be verified or there exist other justifications. They also have the right to object to the processing of their data in certain circumstances, e.g., to stop their PHI from being used for scientific, historical research, or statistical purposes, unless the processing is necessary for the performance of a task carried out in the public interest. Med-health professionals may be able to continue processing if they can show that they have a compelling reason for doing so. Nevertheless, they must tell the data subjects about their right to restrict and object to processing.

The right to data portability

The right to data portability allows patients to obtain and reuse their PHI for their own purposes across a variety of services or products. This new right entitles a patient to obtain from the med-health professionals or institutions, and to transmit to another data handler, e.g., a new doctor or health care center, a copy of his PHI in a structured, commonly used and machine-readable format without undue delay or within a prescribed period of time. It allows them to move, copy, or transfer their PHI easily from one data handler to another in a safe and secure way, without affecting its usability. In the ordinary course of business, doing this would enable the patients to take advantage of services that can use this data to find them a better deal or help them understand their behaviors. No doubt the right only applies to the PHI that the patients have provided to the med-health professionals or institutions.

Data ethics

Laws establish the minimum standards; ethics go well beyond them. There is no dispute that the global privacy regulatory approach has progressed from compliance to accountability, complemented by data governance, and ethics in data processing particularly in cases of AI and automated decisions. Privacy or data laws are always said to be lagging behind innovation, communications, and technology developments, even though most laws are designed to be principles based and technology neutral. Like other data subjects, patients will not be impressed simply by compliance with the laws but by practising ethical standards on which trust and confidence are distinctly based. Data ethical values typically center at fairness, respect, and mutual benefits. In practical terms, they involve genuine choices, meaningful consent, transparency, no bias or discrimination, and fair negotiation or exchange on a level playing field between data handlers and data subjects. It would therefore be crucial if med-health professionals data are also mindful of observing the core ethical principles in formulating their data protection policies and practices.

AI and digital data ethics in med-health sciences and services

Without repeating what is most helpfully articulated in the preceding chapters, AI is often defined to include all systems that display intelligent behavior by analyzing their environment and taking actions—with a significant degree of autonomy—to achieve specific goals.

The merits of deploying AI in med-health sciences and services (including the usage in gastroenterology [6] and ambient intelligence in med-health care) need no elaboration, suffice to say that AI is often regarded by privacy regulators as a high-risk technology creating complex grounds for bias and discrimination posing privacy, security, and other liability concerns. Decisions based on the machine learning algorithms which are probabilistic in nature have given rise to critiques and privacy regulators' raising their eyebrows, particularly because AI systems can perpetuate bias and discrimination on a massive scale originating from raw or generated Big Data.

The use of AI must also comply with the requirements of data protection laws, whether in the form of medical diagnoses, clinical operations, web cookies, behavioral tracking, or profiling.

"Profiling" is defined [7] as "any form of automated processing of personal data consisting of the use personal data to evaluate certain personal aspects relating to a natural person, in particular to analyse or predict aspects concerning that nature person's performance at work, economic situation, health, personal preferences, interests, reliability, behaviour, location or movement."

The law is that a patient has the right not to be subjected to a decision based solely on profiling [8].

Whatever the choices, data handler deploying an AI system will have to be accountable for it. Auditability, which enables the assessment of algorithms, data, and design processes, plays a key role to ensure accountability for AI systems and their outcomes.

Trustworthy ethical AI considerations include human agency and oversight; technical robustness and safety; privacy and data; transparency; diversity, nondiscrimination, and fairness. These core principles are largely shared by regional and international organizations such as EC [9], OECD [10], Global Privacy Assembly [11], and the office of PCPD, Hong Kong [12] (Box 4.2).

From a privacy structure to a privacy culture

Given the established cross-continent or cross-border structures and infrastructures for personal data privacy protection, med-health professionals should be conversant with privacy and confidentiality threats, and what they are required by law, and should ethically do to take preventive, operational, and remedial measures. It should be done not only by holding top-level board meetings at regular intervals but also communications from time to time for protecting privacy on social media and messenger platforms, through emails and in posters around the workplace. In addition to compliance with the laws, med-health professionals should spare no efforts in

BOX 4.2 AI and digital data ethics.

- AI and machine learning algorithms may have high risks in creating and perpetuating bias and discrimination originating from the use of massive raw or generated Big Data, subsequently posing privacy, security, and other liabilities.
- The use of AI (to which med-health institutions are accountable) must also comply with the data protection laws, whether in the form of medical diagnoses, clinical operations, or profiling—including using a patient's personal data to evaluate his health to which he has the right not to be subjected.
- Data ethics (on which patients' trust and confidence are based) complement compliance with the laws especially in cases of AI and automated decisions.
- The core and trustworthy ethical AI principles and considerations—genuine choices, meaningful consent, transparency, no bias or discrimination, and fair negotiation or exchange on a level playing field—should be observed in formulating med-health institutions' data protection policies and practices.

promoting, formulating, and practising core ethical values, together with patients and regulators, for the cultivation, re-enforcement, and engagement of a privacy culture, in-house and global, which would contribute to quality med-health care, confidence, and trust.

References

[1] European Union. Article 9, General data protection regulation; 2023.
[2] Personal Information Protection Law (PIPL), editor. The National People's Congress of the People's Republic of China; 2021.
[3] Creemers R., Webster G. Translation: Personal information protection law of the People's Republic of China 2021. (Accessed 2 April 2023).
[4] Personal Information Protection Law (PIPL), editor. The National People's Congress of the People's Republic of China. Article 49; 2021.
[5] Irish Data Protection Commission. Data protection commission announces decision in Meta (Facebook) inquiry n.d. (Accessed 2 April 2023).
[6] Stewart C, Wong SKY, Sung JJY. Mapping ethico-legal principles for the use of artificial intelligence in gastroenterology. J Gastroenterol Hepatol 2021;36:1143–8. https://doi.org/10.1111/jgh.15521.
[7] European Union. Article 4(4), General data protection regulation; 2023.
[8] European Union. Article 22(1), General data protection regulation; 2023.
[9] High-Level Expert Group on Artificial Intelligence, European Commission. Ethics guidelines for trustworthy AI 2019. (Accessed 2 April 2023).
[10] Organisation for Economic Co-operation and Development. Recommendation of the council on artificial intelligence; 2019.
[11] Global Privacy Assembly. Adopted resolution on accountability in the development and use of artificial intelligence; 2020.
[12] Office of the Privacy Commissioner for Personal Data, Hong Kong. Guidance on the ethical development and use of artificial intelligence; 2021.

CHAPTER 5

Black box medicine

Irwin King[a], Helen Meng[b], and Thomas Y.T. Lam[c]
[a]Department of Computer Science and Engineering, The Chinese University of Hong Kong, Hong Kong Special Administrative Region
[b]Department of System Engineering and Engineering Management, The Chinese University of Hong Kong, Hong Kong Special Administrative Region
[c]The Nethersole School of Nursing, The Chinese University of Hong Kong, Hong Kong Special Administrative Region

Introduction

In recent years, the accelerating advancement of AI technology has dramatically revolutionized healthcare and medicine development. A plethora of cutting-edge innovations is flourishing prosperously, including but not limited to medical imaging diagnosis, robotic surgery, healthcare big data management, and decision support systems. We define AI medicine as software and hardware systems that display intelligent behavior, such as predicting current and future health conditions through data processing [1]. However, decisions made by machine learning algorithms cannot always be explicitly understood or adequately explained to stakeholders, and this is often regarded as "black box medicine." As these tools are increasingly used in healthcare, concerns arise about how various stakeholders, including healthcare professionals, patients, policymakers, researchers, and engineers, incorporate ethics into the AI black box system. Primary concerns include the opaqueness of AI algorithms leading to lack of trust in model predictions, unclear accountability, and muddled responsibility for using AI in healthcare.

Although various AI principles have been proposed and investigated by different organizations [2], we propose to follow these five common and fundamental ethical principles: (1) transparency, (2) fairness and justice, (3) beneficence, (4) responsibility, and (5) privacy. We elaborate on critical issues posed by the ethical and legal use of black box medicine in healthcare and medical applications using five ethical principles with three distinct stages of AI development, namely, data collection, algorithm and model development, and deployment(Figs. 5.1 and 5.2).

Artificial Intelligence in Medicine
https://doi.org/10.1016/B978-0-323-95068-8.00005-4

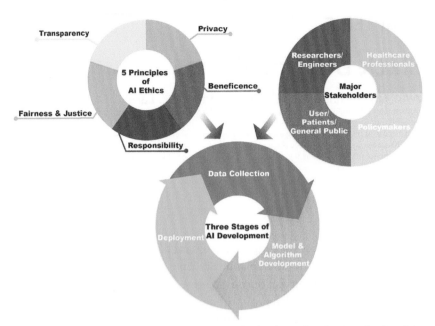

Fig. 5.1 The overview of the AI black box with the five ethical principles involving various stakeholders.

The three stages of AI development with five principles

Typical AI development involves three distinct stages: data collection, model and algorithm development, and deployment. Here, we would like to present how some of these five principles are involved in the first two stages of AI development as the deployment stage is beyond the scope of this chapter.

Data collection

Data collection is an essential step in data-driven AI systems. This AI development stage includes several ethical and regulatory challenges associated with data privacy, fairness, transparency, and conflict of interest [3]. Some of these issues have been discussed in Chapters 2 and 4 on data access, data bias and data equity, and privacy and confidentiality, respectively.

Five Principles of Ethics	THREE STAGES OF AI DEVELOPMENT		
	Stage 1: Data Collection [INPUT]	Stage 2: Models & Algorithm Development [BLACK BOX]	Stage 3: Deployment [OUTCOME]
Transparency	• How is patient data collected, stored, used, and retrieved? • How can data collection and retrieval methods be explained to users? • How can collected data be protected from exposure to third parties?	• AI decision-making models should be made clear. • Patients and users should be aware of how AI decisions are made based on different data sets. • What mechanisms can make AI model operations more transparent? • Design AI models that are explainable and interpretable.	• What external testing measures can ensure the transparent implementation of AI models? • Are there ways to deploy AI models in such a way to limit the trade-off between accuracy and transparency?
Fairness and Justice	• How can data collection and storage systems be designed to ensure that collected data is unbiased, balanced, and sufficient? • Which common biases can be identified at the outset to better inform the data collection process, such as age, sex, or gender?	• How can biases in AI technologies be detected? • Can past data be a good reference for future decisions?	• Are there cultural differences that lead to misinterpretation or incorrect outputs in AI applications? • Policing ensures the fairness and justice of AI model decisions and prevents the possibility of garbage-in and garbage-out scenarios. • What are the most effective evaluation mechanisms for better AI model applications?
Beneficence	• How can patient and public health interests take priority over corporate or profit-making interests in the data collection process? • How can collected data be evaluated in terms of beneficence to patients?	• How can the misuse of AI technology be prevented? • How can AI technology be put to good use without doing any harm to the general public?	• What are the concerns regarding detrimental effects brought about by AI technology when it becomes a necessary evil or a harmful device? • How can we ensure that AI technology is beneficial to users?
Responsibility	• Which parties are responsible for collecting data that is relevant and appropriate to the development of specific AI technologies? • Who is held accountable for evaluating the accuracy and quality of the collected data?	• How can AI models be designed to maintain an appropriate reliance on human input to address responsibility concerns? • How can AI models be designed to provide "traceability" of decision-making processes?	• Who/What is making the decisions? • If AI models make a decision that leads to an undesirable outcome or ethical dilemma, who is accountable? • How can healthcare professionals be trained to incorporate AI technologies appropriately and effectively?
Privacy	• Data privacy must be respected. • How can data be kept confidential throughout the data collection process? • What policies should be formulated for safeguarding the privacy of users' data? • Who has access to collected data?	• How can AI models be designed to pool health data from relevant resources in an ethically acceptable way without sharing patients' records? • How can synthetic data help address data privacy concerns while still creating high-quality AI models?	• How can policymakers create legislation that protects privacy without hindering the development of beneficial AI technologies? • How can surplus data be used for other applications in an ethical way?

Fig. 5.2 Ethical issues of AI for medicine under different stages and five principles of ethics. Showing some key ethical issues with respect to the five ethical principles in the three stages in the development of the AI black box for medicine.

Transparency

The use of "black box" healthcare AIs raises the question of how healthcare professionals can remain **transparent** with patients while working with systems that are not fully transparent [4]. To trust AI models, patients and healthcare professionals need to understand the data sources that led to the development of these models. Patients have strong views about the transparent flow and security of their data used in AI algorithms, e.g., why the data are being collected, how they are being used, how long the data will be stored, etc. A clear acceptance of the AI model and its outcomes requires a transparent mechanism of data collection by healthcare professionals.

With the growing demand for patients' data used in AI models, it becomes policymakers' responsibility to decide when and how patients should be involved in data collection. The standards of data transparency have been provided by the guidance note from the United Nations Development Group addressing data collection with due diligence [5]. AI developers should be sufficiently transparent about the kind of data used and any shortcomings of the software [6]. They should describe what kind of data is collected and why, how it is used and shared with third parties, and what safeguard measures are in place to protect patients' privacy [7]. By implementing aforementioned measures to enhance transparency, AI developers can increase stakeholders' trust in the data collection process and the models' recommendations.

Privacy of data

Collecting data from patients arouses patient privacy concerns. Therefore, it is essential to protect the privacy and avert the unconsented use of patients' data. This has been a long-standing principle in medicine and should remain a key consideration in the development and implementation of all health AIs [4].

From the perspective of patients and the general public, a major concern is whether a large amount of patient data gathered to train AI models is safely protected by the hospitals, research organizations, pharmaceutical companies, insurance companies, and technology companies that store and process the data. Data has become a commodity that is traded and marketed. When healthcare delivery systems expand to become more regional and global, protecting data becomes more difficult, complicated, expensive, and security breaches tend to have more serious impacts [8]. This has significant implications for how data is used by healthcare professionals in practical settings.

When patients receive clinical intervention by AI devices, their generated data is usually automatically collected to further train the AI system to improve its performance in making clinical decisions. Apart from enhancing the performance of the existing AI algorithms, collected data is also useful to develop new AI algorithms to be applied in other clinical settings. For example, many AI polyp detection systems are available on the market that assists endoscopists in detecting polyps automatically and simultaneously during colonoscopy [9]. The AI system stores images of the detected colonic polyps to form a new training dataset to improve polyp detection speed and accuracy. Patients should be aware of the potential purposes of their data when giving informed consent.

The confidential information gathered from each patient to empower AI medicine must be well protected from being used for purposes other than what the patients agreed to. Patients must be well informed about all potential usage of their data and provided an option to choose if they opt-in to contribute their confidential information to a third party, i.e., AI developers, academic research institutes, hospitals, etc. This is important to demonstrate respect for patients' privacy which is an important bioethical principle [10]. For healthcare professionals, their major concern is that biomedical big data may foster a divide between those who accumulate, acquire, analyze, and control such data and those who provide the data but have little control over their use. This is especially true with respect to data collected from underrepresented groups [7]. Policymakers should look into ways to minimize this divide to ensure the secure and fair use of the collected data.

Ethical use of medical data for AI researchers is accomplished by measuring and sharing technical characteristics of the data to preserve patient privacy and meaningful properties of the underlying data and models. In the case of rare diseases, the exchange of medical data from multiple institutions needs specialized AI techniques to safeguard privacy violations [11]. It is possible to carefully design AI models that pool health data from relevant resources in an ethically acceptable way without sharing patients' records and related information. For example, Federated Learning can provide a secure and trustworthy way to collect data without actually sharing the real data [12]. Data augmentation methods using generative AI models by producing realistic but synthetic data can also avoid the long-term dependency on patients' real data. However, the excessive use of data augmentation may consequently hurt more than help the decision process. The lesson learned is that we need to balance out the data-hungry algorithm against the availability of bonafide data.

Models and algorithm development

"Black boxes" can be described as software that is usually designed to help healthcare professionals with patient care, but that does not explain how the input data is analyzed to reach its decision [13]. Machine learning represents the "unknown," "blacked-out," or "mysterious" processes that occur between data input and the derivation of AI. This raises questions about how to communicate possible biases, risks, and error to patients during the informed consent process. AI systems are not totally autonomous. They are fed with human-generated algorithms. Human actions are plagued by errors or influenced by biases. When faulty algorithms are used by big data technology, there is the tendency to reproduce or magnify human errors or biases, resulting in the risk of providing faulty AI related to existing health disparities [8]. Despite these risks, there is tremendous potential for AI technologies to increase the quality and efficacy of healthcare, as long as the development process focuses on patient well-being by implementing the ethical values of transparency, interpretability, beneficence, fairness, and justice.

Transparency

Can healthcare professionals make an important decision without knowing the rationale behind it? Can patients accept treatment without understanding what is happening behind the decision-making process? Can healthcare professionals communicate adequately to patients why they choose a certain treatment and not the other? The lack of trust in machine learning (ML) algorithms poses a significant challenge to the implementation of AI medicine [14].

The key issue is whether the available data and ML models are of sufficient quality and detail to inform the clinical questions of interest. In addition, patients should be made aware of the level of supervision provided by health practitioners over the AI. They should also be informed about whether the use of AI in that particular context is a standard-of-care, experimental, or part of a research project [15].

Healthcare professionals should be frank with patients in ways that they can understand how AI will be used to inform decisions about their care. They should try their best to explain to their patients the purpose of using AI, how it functions, and whether it is explainable. Patients should also be informed about when, if, and how their health data is being retained and fed back into the deep learning of the AI [15]. They should also be transparent about any weaknesses of the AI technology, such as any biases, data breaches, or privacy concerns [7]. Trust and open communication are the keys to facilitating the adoption of AI in medicine.

In a recent systematic review of AI devices in real-life practice, 40% of the applications did not provide a transparent description of the AI system's architecture. The result was a negative influence on the acceptability of their end-users [16]. In another study, only 16% of clinicians perceived an AI-based sepsis prediction system to be helpful. It is difficult for clinicians to trust complex AI algorithms over their own clinical judgment if not provided with explanations to facilitate interpretation of the AI prediction of sepsis. The majority of clinicians requested greater transparency of the AI algorithm because they wanted to know how the AI system predicts sepsis [17]. In dermatology, AI tools have been developed and trained by dermoscopic images to assist in diagnosing skin cancers. A recent systematic review [18] on skin cancer classification using AI technology identified 13 different AI algorithms, in which only one of them was made publicly available for external testing [19].

Trustworthy AI research in healthcare needs transparent AI models, which brings transparent decisions that stakeholders can easily comprehend. Self-explanatory AI models such as decision trees, random forests, and shallow neural networks are less powerful [20]. Complex models such as deep neural networks are higher in accuracy but less transparent. There are ongoing efforts to make high-accuracy AI models more transparent.

Besides AI users and receivers, policymakers need to ensure transparency in any AI model used in clinical practice to avoid inequalities created because of the usage in healthcare among patients from different age groups, gender, ethnicity, or other protected attributes. From the policymakers' point of view, the question arises to what extent the AI-driven solutions approve and demand transparency. What is the ethical solution to providing healthcare professionals with relevant information about the characteristics, features, and underlying assumptions of the AI models? To what extent does the patient have to be made aware that the medical decision support system has input from AI technologies? Whether the healthcare providers must adhere to its recommendations or if they can overrule the machine [21]?

Interpretability

The interpretability of algorithms aims to add human reasoning capabilities to the machine-derived algorithm instead of basing it purely on data. Lack of interpretability in AI models, especially in healthcare, can undermine stakeholders' trust using these models. Interpretable machine learning algorithms allow the end-user to interrogate, understand, debug, and improve the ML system [14].

The brain of the "black box" refers to the machine learning models and algorithms used to perform disease diagnosis and risk prediction based on tasks such as classification, clustering, prediction, etc. The main issue of this "black box" is that it is often difficult to interpret or explain the outcome since the explainability of a "black box" model is often a trade-off between accuracy and interpretability [22]. Often, models with accuracy characteristics are more complex and nontransparent, while interpretable machine learning models (e.g., rule-based systems or decision-trees) are less refined and less accurate. An interpretable model can accurately associate a cause, i.e., feature to the effect, outcome, result, etc., and an explainable model can accurately pinpoint what parameters, i.e., the output of a neural, can justify the results. Hence, machine learning algorithms (e.g., decision trees or linear regression) are interpretable since their parameters can be traced directly to the results. On the other hand, deep learning approaches (e.g., neural networks) are powerful, yet what is being learned in the hidden layers is complex and is difficult to explain.

Currently, there is no known legal framework to encapsulate AI ethics into laws. The interpretability of AI models for medical applications is an essential framework that intends to assist AI developers in sanity checking these models beyond mere performance. The modern AI models are powerful but often intrinsically opaque: they fail to meet the interpretability and transparency of an AI model. This is because interpretability and transparency are often an afterthought in the development process. Looking forward, we need to codify these ethical concepts into designing and implementing AI models and algorithms. For example, additional techniques, such as LIME and SHAP, can be used to regain interpretability when using these models [23].

Beneficence

Beneficence ("Do good") and nonmaleficence ("Do no harm") are two ethical principles and characteristics that the data-driven AI system should exhibit during the designing and development process. Beneficence contributes to the well-being and dignity of the users, while nonmaleficence provides the security and privacy of the data being used by the black box. AI models should be designed in a way that minimizes risks of harm and enhances benefits [7]. Care must be taken to ensure the system reflects important societal values that lead to greater public confidence in the black box and realizable benefits to the community. Although the concept of doing good and not doing harm can be understood intuitively, it could be difficult to implement and deploy due to the differences in policies, frameworks, values and rules that guide the development of AI black box systems.

Fairness and justice

The principle of justice deals with the distribution of resources within a society and the nondiscrimination of individuals. Nonintentional injustice to individuals has become an important issue for AI. Due to the data-driven nature of AI techniques, the selection of datasets for training is a significant source of discrimination [24].

Fairness in machine learning refers to the minimization or elimination of algorithm/model bias due to potential biases in data. The benefits of AI in healthcare may not be evenly distributed. Algorithms may discriminate against specific subgroups for whom representative data is not available. Although no model is perfect, unadjusted algorithmic biases may result in disadvantages for certain groups of people when AI technologies are applied. This could affect patients with rare medical conditions or others who are underrepresented in clinical trials and research [8]. For example, various AI algorithms were developed to help diagnose skin cancer. However, most of the skin lesion images were collected from light-skinned populations from Western countries. It is challenging to apply this AI technology to dark-skinned populations because the algorithm was not yet trained with sufficient dark-skinned images and hence it is hard to guarantee reliable detection for performance in this context [18]. Therefore, greater care must be taken to incorporate data that is both just and fair to improve the efficacy of AI models in treating all members of society.

One of the problematic issues facing engineers in designing and implementing fair systems is that the model is typically designed with existing, but incomplete, data. When missing or new data is introduced, how can we ensure that the system does not react in a biased way against these data? Modifying data features by querying the model with counterfactual examples to ensure the assessment is unbiased is one possible way. By focusing on developing systems that incorporate more complete datasets and also making it possible to introduce new information without sacrificing fairness, we can enhance our capacity to deploy these systems in a way that benefits patients and healthcare professionals alike.

Black box medicine: What is acceptable?

In conclusion, it is not only the clinical performance of health AIs that should be considered when evaluating their effectiveness. Interpretability and explainability of AI algorithms are also important to gain trust of the users and receivers. AI engineers must try hard in developing AI models that actively uphold the five ethical principles to build trust in the data collection, model development, and deployment processes to open the black box and make it understandable to all stakeholders, including patients and healthcare professionals. New approaches using data augmentation and federated learning will be helping us in this endeavor. On the other hand, one could argue that AI-algorithms have to be 100% transparent and interpretable before being applied in patient care. There are many drugs which we have used for many years, yet the underlying pharmacological mechanisms remain elusive. The case of aspirin in cancer prevention, the use of metformin in diabetes control, the use of colchicine in gouty arthritis, just to name a few.

References

[1] Bleher H, Braun M. Diffused responsibility: attributions of responsibility in the use of AI-driven clinical decision support systems. AI Ethics 2022;2:747–61. https://doi.org/10.1007/s43681-022-00135-x.
[2] Zeng Y, Lu E, Huangfu C. Linking artificial intelligence principles; 2018. https://doi.org/10.48550/ARXIV.1812.04814.
[3] Vayena E, Blasimme A, Cohen IG. Machine learning in medicine: addressing ethical challenges. PLoS Med 2018;15, e1002689. https://doi.org/10.1371/journal.pmed.1002689.
[4] Katznelson G, Gerke S. The need for health AI ethics in medical school education. Adv Health Sci Educ 2021;26:1447–58. https://doi.org/10.1007/s10459-021-10040-3.
[5] United Nations Development Group (UNDG). Data privacy, ethics and protection: guidance note on big data for achievement of the 2030 agenda, https://unsdg.un.org/sites/default/files/UNDG_BigData_final_web.pdf. [Accessed 14 March 2022].
[6] Gerke S, Minssen T, Cohen G. Ethical and legal challenges of artificial intelligence-driven healthcare. In: Artificial intelligence in healthcare. Elsevier; 2020. p. 295–336.
[7] World Health Organization. Ethics and governance of artificial intelligence for health. World Health Organization; 2021.
[8] Anom BY. Ethics of Big Data and artificial intelligence in medicine. Ethics Med Public Health 2020;15, 100568. https://doi.org/10.1016/j.jemep.2020.100568.
[9] Hassan C, Spadaccini M, Iannone A, Maselli R, Jovani M, Chandrasekar VT, et al. Performance of artificial intelligence in colonoscopy for adenoma and polyp detection: a systematic review and meta-analysis. Gastrointest Endosc 2021;93:77–85.e6. https://doi.org/10.1016/j.gie.2020.06.059.
[10] Reddy S, Allan S, Coghlan S, Cooper P. A governance model for the application of AI in health care. J Am Med Inform Assoc 2020;27:491–7. https://doi.org/10.1093/jamia/ocz192.

[11] Tom E, Keane PA, Blazes M, Pasquale LR, Chiang MF, Lee AY, et al. Protecting data privacy in the age of AI-enabled ophthalmology. Transl Vis Sci Technol 2020;9:36. https://doi.org/10.1167/tvst.9.2.36.

[12] Jiang M, Wang Z, Dou Q. HarmoFL: harmonizing local and global drifts in federated learning on heterogeneous medical images. Proc AAAI Conf Artif Intell 2022;36:1087–95. https://doi.org/10.1609/aaai.v36i1.19993.

[13] Daniel G, Silcox C, Sharma I, Wright MB. Current state and near-term priorities for AI-enabled diagnostic support software in health care; 2019.

[14] Poon AIF, Sung JJY. Opening the black box of AI-medicine. J Gastroenterol Hepatol 2021;36:581–4. https://doi.org/10.1111/jgh.15384.

[15] Stewart C, Wong SKY, Sung JJY. Mapping ethico-legal principles for the use of artificial intelligence in gastroenterology. J Gastroenterol Hepatol 2021;36:1143–8. https://doi.org/10.1111/jgh.15521.

[16] Yin J, Ngiam KY, Teo HH. Role of artificial intelligence applications in real-life clinical practice: systematic review. J Med Internet Res 2021;23, e25759. https://doi.org/10.2196/25759.

[17] Ginestra JC, Giannini HM, Schweickert WD, Meadows L, Lynch MJ, Pavan K, et al. Clinician perception of a machine learning–based early warning system designed to predict severe sepsis and septic shock*. Crit Care Med 2019;47:1477–84. https://doi.org/10.1097/CCM.0000000000003803.

[18] Brinker TJ, Hekler A, Utikal JS, Grabe N, Schadendorf D, Klode J, et al. Skin cancer classification using convolutional neural networks: systematic review. J Med Internet Res 2018;20, e11936. https://doi.org/10.2196/11936.

[19] Han SS, Kim MS, Lim W, Park GH, Park I, Chang SE. Classification of the clinical images for benign and malignant cutaneous tumors using a deep learning algorithm. J Investig Dermatol 2018;138:1529–38. https://doi.org/10.1016/j.jid.2018.01.028.

[20] Quinn TP, Jacobs S, Senadeera M, Le V, Coghlan S. The three ghosts of medical AI: can the black-box present deliver? Artif Intell Med 2022;124, 102158. https://doi.org/10.1016/j.artmed.2021.102158.

[21] Amann J, Blasimme A, Vayena E, Frey D, Madai VI. Explainability for artificial intelligence in healthcare: a multidisciplinary perspective. BMC Med Inform Decis Mak 2020;20:310. https://doi.org/10.1186/s12911-020-01332-6.

[22] Johansson U, Sönströd C, Norinder U, Boström H. Trade-off between accuracy and interpretability for predictive in silico modeling. Future Med Chem 2011;3:647–63. https://doi.org/10.4155/fmc.11.23.

[23] ElShawi R, Sherif Y, Al-Mallah M, Sakr S. Interpretability in healthcare: a comparative study of local machine learning interpretability techniques. Comput Intell 2021;37:1633–50. https://doi.org/10.1111/coin.12410.

[24] Beil M, Proft I, van Heerden D, Sviri S, van Heerden PV. Ethical considerations about artificial intelligence for prognostication in intensive care. Intensive Care Med Exp 2019;7:70. https://doi.org/10.1186/s40635-019-0286-6.

CHAPTER 6

Clinical evidence

Kendall Ho[a], Sarah Park[a], Michael Lai[b], and Simon Krakovsky[b]
[a]University of British Columbia Faculty of Medicine, Emergency Medicine, Vancouver, BC, Canada
[b]University of British Columbia, Vancouver, BC, Canada

Introduction

Healthcare incessantly uses data for historical documentation and present-day diagnosis, yet the real-time extraction of data is underutilized to improve future patterns of care. Meanwhile, other nonmedical disciplines use real-time data in a variety of ways: business intelligence in banking or investing, projection of inventory for just-in-time supply chain management, or autonomous driving to improve traffic flow and optimize safety.

With progress in digital technologies such as cloud computing, personal digital devices, individual sensors, and data science techniques, the opportunity of capturing patients' data to support health and wellness today and using the data for future machine learning and artificial intelligence is unprecedented. Healthcare needs to learn from other disciplines and paradigms to improve healthcare delivery.

This chapter dissects some of the principles and practices that underpin the use of artificial intelligence (AI) in clinical medicine and provides several examples of clinical applications to provide readers with insights to follow its development going forward.

What is AI in clinical medicine?

The ongoing integration of AI into clinical medicine is bred out of both necessity and desire. The COVID-19 pandemic is a prime example, as it forced many health-related services to adapt to remote delivery through telemedicine [1]. Similar pressures are being applied to the healthcare sector as they attempt to digitize services and manage the quantity and complexity of data in their purview. The impetus for AI stems from the healthcare industry's focus on continuous quality improvement (CQI), integrated health systems, extracting knowledge from data, digital public health surveillance, personalized medicine, and more. For example, CQI has become a standardized protocol for regularly evaluating and improving policies,

Artificial Intelligence in Medicine
https://doi.org/10.1016/B978-0-323-95068-8.00006-6

protocols, and performance in both the clinical and administrative settings, paving the way for new technology implementation [2]. However, in order to have a meaningful discussion on the merits and shortcomings of AI, it is first valuable to contextualize its need by discussing existing limitations on the status quo.

The current medical system is nearing its breaking point, pushed to its limit by an aging population, rising healthcare demand, and resource shortages. The aging population in developed countries around the world is driving the demand for precious healthcare resources, as they live longer while needing consistent care for one or many chronic illnesses [3]. The ripple effects are highlighted in examples such as the projected soaring demand for new general surgeons to manage future cancer care patients while the current population-based ratios for general surgeons remain relatively unadjusted [4]. The impact of insufficient general surgeons means that wait times, complications for delayed treatments, and premature death will increase. Another example is the 18.4% increase in emergency department visits in the United States from 2006 to 2014, or from 89.6 to 106.0 million [5]. Similarly, over a third of physicians are over 55 years old, and the same fraction is expected to retire within the decade [6]. Meanwhile, healthcare workers experience tremendous work stressors and staff burnout is further contributing to this problem [7]. From a taxpayer's perspective, physician burnout alone costs the United States between $2.6 and $6.3 billion dollars per year [8]. For the resilient care providers remaining, a randomized controlled trial of attending physicians increasing supervision of residents did not significantly decrease physician error [9].

These examples point to a gap at both an institutional and individual level, underscoring an imperfect healthcare system as well as an imperfect workforce. AI is well-positioned to help fill the gap at both the systemic and worker level. AI can serve as a reliable interface between the growing complexity and quantity of health-related data and the overburdened staff to ultimately improve patient clinical outcomes. Taking a step back, AI can be defined as "the ability of a digital computer or computer-controlled robot to perform tasks commonly associated with intelligent beings" [10]. The aspiration is that AI should be able to reason, learn, and apply knowledge, which are the hallmarks of human intelligence. The current applications for AI in clinical medicine are already numerous and growing, ranging from robot-assisted surgeries to advanced electronic health records which help staff coordinate and streamline care. Yet in many instances, AI in clinical care is only at the starting phase of adoption but not in any way close to mainstream. Ultimately, AI in clinical medicine is an evolving technological strategy, driven by human innovation and institutional needs.

Technology has been in constant evolution for hundreds of years to aid researchers and healers in improving patient care and patient outcomes. AI is plainly yet another medium to further this goal. From the invention of the stethoscope used for listening to heart and lung sounds to the discovery of radiography to noninvasively see into the body, clinical medicine is being nourished with new avenues for valuable information. However, as these technologies evolve, the ways in which we collect, store, access, and interpret this data are becoming increasingly complex. This is where AI, through areas like machine learning and Natural Language Processing (NLP), facilitates our ability to keep up with the technology by acting as the interface between highly skilled staff and data.

Case examples

An increasing number of clinical domains using machine learning and artificial intelligence in clinical decision-making, diagnosis, and management are rapidly emerging. Here, we have selected a few scenarios in different clinical specialties to illustrate some of the leading examples.

Chronic disease management

Patients with chronic diseases face significant challenges in managing their illnesses and are highly vulnerable to recurrent hospitalizations. For example, heart failure is one of the leading causes of morbidity and mortality in Canada, requiring a high level of healthcare utilization [11,12]. Older adults are at much higher risk of decompensation, requiring frequent emergency department visits and hospitalizations with a decreased level of quality of life as a result [13]. For example, approximately 1 in 5 older adults require re-hospitalizations within 30 days of discharge and 1 in 3 within 3 months [14]. Similarly, 21% of HF patients are readmitted to a hospital within 1 month [15]. Factors that need to be considered to improve these patients' quality of life and wellness include but are not limited to having insufficient and complex information about co-morbidities, self-management, and health professional-mediated approaches to improve care or detect deterioration, poorly defined responsibilities and coordination of different health professionals, and patient and caregiver barriers to health information complexity to achieve sound health literacy for optimal self-management [16,17].

Close monitoring of these patients with multidisciplinary support together with optimal self-management is seen as viable ways to help these individuals to maintain wellness and reduce healthcare utilization [18]. Studies on the use of home monitoring devices appeared in the literature over the last decade, with mixed clinical outcomes in terms of the benefits of this approach to improve clinical care of patients [19]. However, these studies did not look at the utility of the data of these patients beyond the current clinical episodes. Literature starts to emerge with the use of the cumulative data of the patients to not only manage the current episodes to diagnose illnesses, but also how to use AI to start to carry out early detection for expeditious intervention before further deterioration requiring emergent care, future prediction of disease progression, and optimal self-management for calibration. Cross application of this approach from one chronic disease to another, for example, TEC4Home, sees the applicability of this foundational approach from heart failure to be extended to hypertension, polypharmacy, and multimorbidity [20,21] (Box 6.1).

Radiology

When lay people ponder the applications for artificial intelligence in clinical medicine, many jump to assume that radiologists will quickly be out of a job

---

BOX 6.1 Example 1. Heart failure: TEC4Home [22].

To support patients during their home convalescence from heart failure (HF) exacerbation after hospital discharge, TEC4Home project provided them with a set of home health monitoring equipment for 60 days to track daily blood pressure, heart rate, weight, oxygen saturation, and subjective symptoms such as shortness of breath. Patients were given instructions to look for abnormal measurements or symptoms of deterioration, and a nurse remotely monitored patient data daily for abnormalities and contacted patients as needed. Our feasibility study with 70 patients showed a 79% reduction of emergency department (ED) revisits, an 87% reduction in hospital readmissions, an 58% decrease in health system costs of management, and a 19% increase in patients' general quality of life score. We have since completed an additional 270 patients, and we are actively mining the physiologic data using machine learning protocol to correlate with patient symptoms, ED visits and rehospitalization so as to look for changes that can predict impending ED visits. This AI approach strives to improve the identification of patients benefiting from early home interventions to avoid deteriorations to the point of needing ED visits.

due to deep learning technology. While human radiologists have not been replaced by artificial intelligence as of now, there have been tremendous advancements over the past few decades. In fact, high demand and rapidly advancing technology have put added pressure on hospital administration and radiology staff to optimize practices. Efficiency is being promoted as radiologists have increased their interpretation from 2.9 to 16.1 images per minute from 1999 to 2010, respectively [23]. The ramifications on effectiveness must also be considered as this statistic implies that some radiologists now interpret one image every 3–4 s. Machine learning is ensuring robust effectiveness by assisting radiology staff.

This new area of medicine is called radiomics, extracting vast amounts of data about image features through data-characterization algorithms [24]. This allows machines to develop the capacity to appreciate variations in medical images and subsequently interpret and provide quantitative clinical information. This information can manifest as the quantification of tumor phenotype or decision support tools for staff in addition to being able to estimate cancer-specific mortality [25]. For lymph node metastasis, artificial intelligence has a higher sensitivity than humans with only a slightly lower specificity [26]. While the technology is enabling optimized care to compensate for the increase in demand, it is not yet at the level to completely replace highly-skilled staff [27]. However, the advantage of deep learning algorithms is that they can learn and thus are able to improve as more data is fed into it. With time, it is likely that advances will demonstrate both superior sensitivity and specificity over humans. For now, there is still plenty of work for radiologists. In fact, they are working in robust cooperation with technology to streamline workflow and improve patient care (Box 6.2).

Mental health and psychiatry

AI in medicine is rapidly evolving in the domain of mental wellness and psychiatry [29]. As the thoughts and moods of individuals are most often manifested through body language, speech, and behavior, digital detection and recording of these areas constitute a rapid area of development in mental health.

The National Institute of Mental Health has introduced the Research Domain Criteria as a research framework to capture new ways to investigate and evaluate mental disorders, from genomics to novel behavioral measures and self-reports. In this context, there is a variety of tools to capture individuals' characteristics and changes digitally via smartphones and connected

BOX 6.2 Example 2. Fractures: AI diagnosis using deep convolutional neural network [28].

This study published in the European Journal of Radiology demonstrates the potential of employing Deep Convolutional Neural Network (DCNN) to help radiologists spot fractures that have not been apparent to the human reader. Hip fractures are a common cause of hospitalization in the elderly, and most are diagnosed with X-ray. But some patients have fractures that are hidden on these exams, perhaps due to overlying soft tissues or other technical factors. In fact, the rate of hidden hip fractures on X-ray is estimated to be between 3% and 10% of all negative hip or pelvic exams taken for trauma. Seven radiologists read hip X-rays with and without the use of the AI algorithm. The study included 327 patients who underwent pelvic CT or MRI and were diagnosed with hip fractures; the AI algorithm for X-ray was trained with 302 of these exams, while the remaining 25 cases were used for testing the algorithm. As a result, the radiologists could spot 83% of the fractures. DCNN's accuracy reached 91%. This has led the FDA to start allowing AI algorithms for clinical decision support as of 2018.

devices—digital phenotyping [30]. An example of this digital phenotyping for assessment is ecological momentary testing, where digital sensors and smartphones collect individuals' physiologic data and frequency of smartphone usage such as phone calls or texting, and these streams of data are aggregated to a unified data platform to be compared to the individuals' own electronic diaries for analysis of mental wellness or manifestations of diseases [31]. There are also different ways for individuals to capture their own moods and emotions [32]. Furthermore, using Natural Language Processing (NLP) as a way to analyze speeches or texts is another way to capture individuals' sentiments, thought content and thought organization, helping to detect any mood or thought disorders [33] (Box 6.3). Use of NLP to identify those at risk of suicide in a social media chat group [34], or screening for those at risk of suicidal behavior after discharge through patients' electronic health records [35] are two such examples. Chatbot, virtual reality, and augmented reality further contribute to the emergence of digital therapeutics, where promising treatments of mental health illness digitally without medications could be useful [36]; for example, the use of a fully automated conversational agent as a type of cognitive behavioral therapy for patients with anxiety or depression [37], or use of virtual reality for the treatment of agoraphobia [38].

> **BOX 6.3 Example 3. Suicidal risk: Electronic health records (EHR) predictions [35].**
>
> This study illuminated data fields in a large amount of longitudinal historical data available in electronic health records (EHR) that could predict patients' future risk of suicidal behavior. The researchers looked at 15 years of inpatient and outpatient visits and identified 1,728,549 patients with three or more visits. They identified a list of ICD-9 codes that best identified suicide attempt cases and developed models for men and women to correlate a list of data fields with suicidal outcomes in a subset of data. They then validated the models with a new subset of data and calculated the overall cumulative risk scores. The model achieved a 33%–45% sensitivity and a 90%–95% specificity in being able to predict patients' future suicidal behavior 3–4 years in advance. Common risk factors, such as substance abuse or psychiatric disorders, together with less conventional ones, such as chronic conditions or certain injuries, were highlighted. This approach allowed health organizations to leverage machine learning to apply to existing large amount of health data to screen for future risks of suicidality.

Opportunities of using these digital approaches singularly or in combination allow for digital detection of individuals' sentiments to support and augment psychological and psychiatric detection, diagnosis, treatment, and maintenance of wellness for patients. With the capturing of all these data, the use of AI to apply screening, diagnosis, treatment, and prediction of mental health issues or psychiatric illnesses become possible at both the individual and population health levels. The quest is how to further demonstrate evidence and scale up effective digital psychiatry for future utilization and refinement [39].

Cardiology

Atrial fibrillation (AF) is a common condition that increases the risk of stroke, heart failure, and emergency department visits. Oral anticoagulation has proven to be effective in reducing the risk of stroke in patients diagnosed with AF. However, AF is often paroxysmal or asymptomatic, making it difficult to diagnose [40]. Despite the clear benefits of screening, current hardware advancements alone (such as smartwatch ECGs) has shown to be of little yield in identifying silent AF [41] (Box 6.4).

> ## BOX 6.4 Example 4. Atrial fibrillation (AF): AI algorithm for ECGs.
> To better identify select populations for AF surveillance, an AI-enabled ECG algorithm was developed to identify silent AF based on a standard 12-lead ECG obtained during sinus rhythm [42]. Patients with at least one ECG showing AF within 31 days after the sinus-rhythm ECG were classified as being positive for AF. The algorithm was developed using half a million ECGs from over 100,000 patients in a single US center, demonstrating a sensitivity, specificity, and accuracy of 80%. Furthermore, it could be applied retroactively to previous digitally stored ECGs. This may help stratify patients for prospective AF testing with Holter monitor, or by setting an even higher detection threshold, help clinicians decide to initiate anticoagulation for patients with imminent AF and high risk of stroke [40].

Robotics and rehabilitation

Impaired mobility has been associated with poorer quality of life [43]. Therefore, many individuals requiring rehabilitation have improving walking function as an important goal. Gait-related impairments may be due to a wide range of causes, such as neurological (e.g., stroke) and nonneurological conditions (e.g., degenerative joint disease) [44]. Rehabilitation commonly focuses on practicing walking movements repetitively to induce improvements and strengthen muscle capability and improve coordination [45]. While effective, there are issues with regard to maintaining consistency of exercises and increasing volume and intensity of exercises due to the overburden of the physiotherapists [46]. Furthermore, reducing the amount of burden on the therapists may reduce healthcare costs by reducing the amount of direct supervision time by the therapist [47].

One possible solution is the use of robotics in rehabilitation. For lower limb rehabilitation, robotic systems can be categorized into stationary systems, overground walking systems, and wearable robotic walking devices [46]. The rate of development in this field is exponentially increasing with a trend toward specializing target of these robotic interventions [48]. The current research shows modestly positive benefits for a variety of conditions, such as stroke [49], spinal cord injury [50], and cerebral palsy [51] (Box 6.5).

Rehabilitation has been shown to show maximum benefits when intense and repetitive [53]. Thus, these rehabilitation robots can provide passive movement, support the weight of patients, and are able to increase the duration, variety, and quality of the rehabilitation exercises based on the patient's state [54]. While most of the current research is focusing on neurological

> **BOX 6.5 Example 5. Gait impairment: Personalized and optimized dose of rehabilitation [52].**
> Zeng (2021) identified a novel gait detection device integrated with smart sensor shoes as part of a lower limb exoskeleton robot. Through a new algorithm model of BP neural network based on support vector machines (SVMBP), the researchers identified that this combined model could obtain high accuracy of human gait detection with an accuracy of 97.5% in both classification and recognition. Gait recognition is fundamental for subsequent gait planning and control; therefore, this algorithm-enhanced sensor can be used in the future for robotic interventions to automatically adjust accordingly to the gait characteristics and provide necessary support to the lower limb joints in the walking phases to individualize rehabilitation appropriately.

impairments, there are opportunities for robotics to augment rehabilitation services regardless of impairments to allow for maximal in-person care by physical therapists and increase the volume of repetitive task-based therapy to ultimately improve walking function and quality of life.

What is needed for AI?

Clinical medicine has seen "AI winters" of the past, periods of great hype and enthusiasm followed by bouts of low funding and profound disillusionment when faced with the reality of implementation and integration. However, with advancing technology to capture and process ever-increasing amounts of data, a multidisciplinary approach, and emerging frameworks to guide implementation and use, there may be a sufficient and favorable environment for AI to bloom now.

Technology has progressed a long way from when AI was first envisioned in clinical medicine. Our ability to collect data via sensors has only grown more efficient and comprehensive. Traditional biomarkers such as the electrocardiogram (ECG) have long provided significant clinical value at an affordable cost, while also providing easy data to collect and analyze. As such, they are one of the premier tools to be complemented and assisted by AI [40]. However, advancements in technology, particularly mobile and wearable sensors, have also expanded the potential for novel vital signs such as heart rate variability and galvanic skin response to offer diagnostic and prognostic value [55,56]. Continuous remote monitoring of these bio-markers can yield rich clinical data sets for individual patients, for which we now have the capacity to store and process this at an affordable cost.

With the advent of increased access to technology has also come a reduction in costs. The parts needed to develop a deep learning model, such as a graphics processing unit, have become considerably more affordable. A modern graphics processing unit (GPU) may cost roughly $1000, with the ability to process 3000 images per second depending on the machine learning model. Once trained, it is capable of processing 260 million images in 1 day, as machines can work without sleep, at the cost of an average laptop [57]. Technology has evolved at an unprecedented rate, paving the way for increasing usage and implementation of AI.

One thing that we need is to produce evidence that AI-tools can improve clinical outcome of patients and it is superior to the existing model of care using human judgment, diagnosis, and management. Important clinical outcomes include reduced hospital admission, reduced ICU care, and ultimately reduced mortality. However, despite explosion of medical literature on ad hoc use of AI devices, existing evidence is small scale, nonrandomized studies using outcome parameters that do not necessarily indicate clinical superiority that benefit patients. A recent systematic review found only 39 randomized controlled trials to date comparing AI-assisted tools to standard of care management [58] (Fig. 6.1). While two-thirds of the studies demonstrated improvement in primary or secondary endpoints compared to standard of care, these endpoints were not particularly clinically relevant, i.e., leading to change in management and reducing hospital admission or mortality. To illustrate this point, nine of the studies examined AI-assisted adenoma detection systems during colonoscopy. All successfully assisted endoscopists to detect more adenomas during colonoscopy, yet only one study was able to demonstrate improved detection of adenomas of all sizes. The other studies could only improve detection of diminutive or small adenomas. As suggested by the US Multi-Society Task Force on Colorectal Cancer, patients with 1–2 nonadvanced diminutive or small adenomas are low risk and can be followed up with surveillance colonoscopy in 7–10 years. Thus, the clinical value of improved small adenoma detection is unclear at this point. Furthermore, AI did not increase advanced adenoma or colorectal cancer detection in these studies, and the only single-reported long-term outcome was in-hospital mortality. Future studies need to emphasize the impact of AI-assisted tools on long-term clinical outcomes.

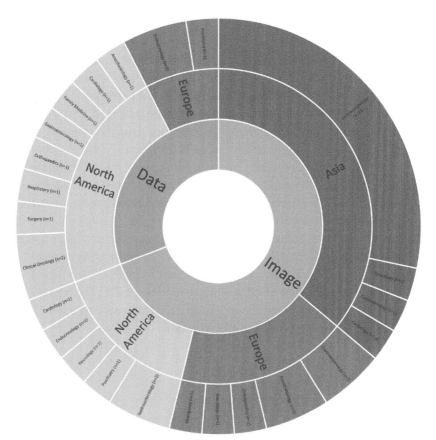

Fig. 6.1 Distribution of publication origin and specialty [58].

Barriers of AI

Despite the technical developments, there are several barriers such as AI literacy within the workforce, AI-specific regulations, and policies and frameworks to guide ethical usage. Despite the exponential increase in research development and uptake of AI in the past few decades, there is still lots of work to be done as these challenges are important to acknowledge and address to move forward.

Adoption and scaling

Even if the technology is fully developed and is backed with evidence, if the stakeholders are not embracing adoption, there will be challenges in the implementation of AI. There are several factors that lead to the adoption of AI as a challenge for healthcare. Primarily, there is a lack of understanding or training of AI for healthcare professionals to critically appraise and adopt AI in clinical practice. Due to the lack of training and a steep learning curve, health-care professionals are less likely to adopt AI into practice. Second, due to the heterogeneity of the healthcare systems, there would be challenges in scaling up AI as each system would require a different onboarding system to integrate AI into the workflow. Having to validate and adapt AI applications to specific contexts can require significant expenditures [59], and regular maintenance and updates to constantly evolve with changes in the real world. Third, studies looking at AI in clinical practice are developed retrospectively and do not cre-ate the same results in other datasets with different patient demographics [60]. To best address these issues for adopting and scaling up AI, there is a need for unified data collection across healthcare systems, comprehensive training of AI for healthcare professionals [61], and continuous evaluation of AI by technical staff and researchers as clinical practice continues to evolve. With improved training and understanding of AI, healthcare professionals will be able to crit-ically appraise and question erroneous results [62] that may occur due to AI to ultimately better integrate AI into the workflow.

Clinical evidence

Evidence-based practice is irrefutably important, especially in AI. However, there is a clear lack of transparency in the decision-making process, robust methodologies, and AI development in peer-reviewed journals. Many tech-niques of AI, such as artificial neural networks or deep learning are known as a "black box" due to the opaque nature of its' inner workings. Thus, when AI is used to make important decisions for patients, it is difficult to explain what was considered in the decision-making process in an understandable way [63]. Current research that compares the efficiency of AI with clinicians has been found to be unreliable and lacks primary replication [61]. Further-more, randomized controlled trials (RCTs) are the "gold standard" for research, but most studies are feasibility or pilot trials which are unable to provide robust evidence on the effects of AI. Two systematic reviews have shown that there are very few RCTs with clinically meaningful outcomes or full adherence to reporting standards [58,64]. There is also a lack of algo-rithm development details due to issues with intellectual properties and a lack of peer-reviewing. To address these issues, there needs to be more work done in increasing adherence to reporting guidelines specific to AI (e.g., TRIPOD) increasing awareness and opportunity for peer-review and using more rigorous research design to assess the true effect, both benefits and drawbacks, of using evidence-based AI in healthcare.

Future of AI

Much research is now actively being done on artificial intelligence, and this field is still very much in its infancy at this time and requires rigor in carrying out clinical research [65,66]. It is clear that AI carries great promise for the future in clinical medicine in three complementary and important ways.

- *Applied AI*: Augmentation for health professionals to improve capacity in primary care, such as in low resource regions where health professionals are in shortage so patients can be screened first prior to accessing healthcare, or extending screening of diseases and bringing specialists' expertise in primary care to identify patients at risk or with illnesses for early intervention [67]. The voluminous amount of data also provides the opportunity to correlate different datasets to find novel linkages for detection, causes, and management of diseases, such as relating consumer data from fitness trackers to individuals' electronic health records to detect sepsis or manage wellness [68]. Expanding on these known AI approaches in a variety of disease entities to demonstrate efficacy will be important.
- *Implementation science of AI*: Change management and introduction of innovative medical approaches take time and require an understanding of human behavior and ethics that govern them. Scaling up AI necessitates thoughtful consideration of ethical issues, and how to guide its sensible introduction into healthcare [69]. This includes but are not limited to the consideration of ethics, culture, health policies, and how human resourcing in healthcare will be altered [65,67,69]
- *Science of AI*: Current state-of-the-art artificial intelligence approaches will no doubt lead to the future evolution of this rapidly evolving field of computer and data science [70]. With the rapid introduction of technological changes—such as 5G, quantum computing, and edge computing—new and better approaches will no doubt contribute to the science of AI developments [71,72].

Interested readers can find deeper discussion of this topic in other chapters in this book.

Conclusion

The future of artificial intelligence in clinical medicine is bright, as government regulators, administrators, healthcare workers, and other key

BOX 6.6 Main takeaways.

- AI can be applied in many medical disciplines to clinically support individual patient care and also improve population health management, stratification and prediction.
- While there is an increasing number of publications in the literature on AI in clinical medicine, more rigorous research studies and higher quality of evidence are needed to validate AI in improving clinical outcomes compared to conventional approaches.
- Advancement of AI in medicine will be in three key directions: applied AI to use known methodologies to interrogate existing data to improve clinical care; AI implementation science AI to integrate AI into clinical management protocols and scale up and spread; and AI science to conduct research on novel data science methodologies to scrutinize health data in unprecedented ways.

stakeholders pay more attention to the need to leverage this technology in order to improve patient care outcomes. The health sector is emerging to embrace AI in clinical care, and future clinical research and implementation science need to focus on the evidence of AI in truly improving care of patients, improving cost effectiveness of healthcare delivery, and increasing health capacity to bring benefits to more patients locally and globally. AI will no doubt transform clinical medicine as it will be practiced tomorrow (Box 6.6).

References

[1] Moazzami B, Razavi-Khorasani N, Dooghaie Moghadam A, Farokhi E, Rezaei N. COVID-19 and telemedicine: immediate action required for maintaining healthcare providers well-being. J Clin Virol 2020;126, 104345. https://doi.org/10.1016/j. jcv.2020.104345.

[2] Antony J, Palsuk P, Gupta S, Mishra D, Barach P. Six Sigma in healthcare: a systematic review of the literature. Int J Qual Reliab Manag 2018;35:1075–92. https://doi.org/10.1108/IJQRM-02-2017-0027.

[3] Harper S. Economic and social implications of aging societies. Science 2014;346:587–91. https://doi.org/10.1126/science.1254405.

[4] Ellison EC, Pawlik TM, Way DP, Satiani B, Williams TE. The impact of the aging population and incidence of cancer on future projections of general surgical workforce needs. Surgery 2018;163:553–9. https://doi.org/10.1016/j.surg.2017.09.035.

[5] Lin MP, Baker O, Richardson LD, Schuur JD. Trends in emergency department visits and admission rates among US acute care hospitals. JAMA Intern Med 2018;178:1708. https://doi.org/10.1001/jamainternmed.2018.4725.

[6] Markit I. The complexities of physician supply and demand: Projections from 2015 to 2030; 2017.

[7] Tawfik DS, Profit J, Morgenthaler TI, Satele DV, Sinsky CA, Dyrbye LN, et al. Physician burnout, well-being, and work unit safety grades in relationship to reported medical errors. Mayo Clin Proc 2018;93:1571–80. https://doi.org/10.1016/j.mayocp.2018.05.014.

[8] Han S, Shanafelt TD, Sinsky CA, Awad KM, Dyrbye LN, Fiscus LC, et al. Estimating the attributable cost of physician burnout in the United States. Ann Intern Med 2019;170:784. https://doi.org/10.7326/M18-1422.

[9] Finn KM, Metlay JP, Chang Y, Nagarur A, Yang S, Landrigan CP, et al. Effect of increased inpatient attending physician supervision on medical errors, patient safety, and resident education: a randomized clinical trial. JAMA Intern Med 2018;178:952. https://doi.org/10.1001/jamainternmed.2018.1244.

[10] Copeland BJ. Artificial intelligence: A philosophical introduction. Oxford, UK ; Cambridge, MA: Blackwell; 1993.

[11] Tran DT, Ohinmaa A, Thanh NX, Howlett JG, Ezekowitz JA, McAlister FA, et al. The current and future financial burden of hospital admissions for heart failure in Canada: a cost analysis. CMAJ Open 2016;4:E365–70. https://doi.org/10.9778/cmajo.20150130.

[12] Virani SA, Bains M, Code J, Ducharme A, Harkness K, Howlett JG, et al. The need for heart failure advocacy in Canada. Can J Cardiol 2017;33:1450–4. https://doi.org/10.1016/j.cjca.2017.08.024.

[13] van Riet EES, Hoes AW, Limburg A, Landman MAJ, van der Hoeven H, Rutten FH. Prevalence of unrecognized heart failure in older persons with shortness of breath on exertion: unrecognized HF in older persons with shortness of breath on exertion. Eur J Heart Fail 2014;16:772–7. https://doi.org/10.1002/ejhf.110.

[14] Jencks SF, Williams MV, Coleman EA. Rehospitalizations among patients in the medicare fee-for-service program. N Engl J Med 2009;360:1418–28. https://doi.org/10.1056/NEJMsa0803563.

[15] Krumholz HM, Merrill AR, Schone EM, Schreiner GC, Chen J, Bradley EH, et al. Patterns of hospital performance in acute myocardial infarction and heart failure 30-day mortality and readmission. Circ Cardiovasc Qual Outcomes 2009;2:407–13. https://doi.org/10.1161/CIRCOUTCOMES.109.883256.

[16] Hayes SM, Peloquin S, Howlett JG, Harkness K, Giannetti N, Rancourt C, et al. A qualitative study of the current state of heart failure community care in Canada: what can we learn for the future? BMC Health Serv Res 2015;15:290. https://doi.org/10.1186/s12913-015-0955-4.

[17] Rustad E, Furnes B, Cronfalk BS, Dysvik E. Older patients' experiences during care transition. Patient Prefer Adherence 2016;769. https://doi.org/10.2147/PPA.S97570.

[18] Díez-Villanueva P, Alfonso F. Heart failure in the elderly. J Geriatr Cardiol 2016;13:115–7. https://doi.org/10.11909/j.issn.1671-5411.2016.02.009.

[19] Liu L, Stroulia E, Nikolaidis I, Miguel-Cruz A, Rios RA. Smart homes and home health monitoring technologies for older adults: a systematic review. Int J Med Inform 2016;91:44–59. https://doi.org/10.1016/j.ijmedinf.2016.04.007.

[20] Krueger K, Botermann L, Schorr SG, Griese-Mammen N, Laufs U, Schulz M. Age-related medication adherence in patients with chronic heart failure: a systematic literature review. Int J Cardiol 2015;184:728–35. https://doi.org/10.1016/j.ijcard.2015.03.042.

[21] Omboni S, McManus RJ, Bosworth HB, Chappell LC, Green BB, Kario K, et al. Evidence and recommendations on the use of telemedicine for the management of arterial hypertension: an international expert position paper. Hypertension 2020;76:1368–83. https://doi.org/10.1161/HYPERTENSIONAHA.120.15873.

[22] Ho K, Novak Lauscher H, Cordeiro J, Hawkins N, Scheuermeyer F, Mitton C, et al. Testing the feasibility of sensor-based home health monitoring (TEC4Home) to support the convalescence of patients with heart failure: pre–post study. JMIR Form Res 2021;5, e24509. https://doi.org/10.2196/24509.

[23] McDonald RJ, Schwartz KM, Eckel LJ, Diehn FE, Hunt CH, Bartholmai BJ, et al. The effects of changes in utilization and technological advancements of cross-sectional imaging on radiologist workload. Acad Radiol 2015;22:1191–8. https://doi.org/10.1016/j.acra.2015.05.007.

[24] Gillies RJ, Kinahan PE, Hricak H. Radiomics: images are more than pictures, they are data. Radiology 2016;278:563–77. https://doi.org/10.1148/radiol.2015151169.

[25] Tang A, Tam R, Cadrin-Chênevert A, Guest W, Chong J, Barfett J, et al. Canadian association of radiologists white paper on artificial intelligence in radiology. Can Assoc Radiol J 2018;69:120–35. https://doi.org/10.1016/j.carj.2018.02.002.

[26] Wang H, Zhou Z, Li Y, Chen Z, Lu P, Wang W, et al. Comparison of machine learning methods for classifying mediastinal lymph node metastasis of non-small cell lung cancer from 18F-FDG PET/CT images. EJNMMI Res 2017;7:11. https://doi.org/10.1186/s13550-017-0260-9.

[27] Hosny A, Parmar C, Quackenbush J, Schwartz LH, Aerts HJWL. Artificial intelligence in radiology. Nat Rev Cancer 2018;18:500–10. https://doi.org/10.1038/s41568-018-0016-5.

[28] Mawatari T, Hayashida Y, Katsuragawa S, Yoshimatsu Y, Hamamura T, Anai K, et al. The effect of deep convolutional neural networks on radiologists' performance in the detection of hip fractures on digital pelvic radiographs. Eur J Radiol 2020;130:109188. https://doi.org/10.1016/j.ejrad.2020.109188.

[29] Shore JH, Schneck CD, Mishkind MC. Telepsychiatry and the coronavirus disease 2019 pandemic—current and future outcomes of the rapid virtualization of psychiatric care. JAMA Psychiatry 2020;77:1211. https://doi.org/10.1001/jamapsychiatry.2020.1643.

[30] Torous J, Onnela J-P, Keshavan M. New dimensions and new tools to realize the potential of RDoC: digital phenotyping via smartphones and connected devices. Transl Psychiatry 2017;7:e1053. https://doi.org/10.1038/tp.2017.25.

[31] Parrish EM, Kamarsu S, Harvey PD, Pinkham A, Depp CA, Moore RC. Remote ecological momentary testing of learning and memory in adults with serious mental illness. Schizophr Bull 2021;47:740–50. https://doi.org/10.1093/schbul/sbaa172.

[32] Sharon T. Self-tracking for health and the quantified self: re-articulating autonomy, solidarity, and authenticity in an age of personalized healthcare. Philos Technol 2017;30:93–121. https://doi.org/10.1007/s13347-016-0215-5.

[33] Torous J, Bucci S, Bell IH, Kessing LV, Faurholt-Jepsen M, Whelan P, et al. The growing field of digital psychiatry: current evidence and the future of apps, social media, chatbots, and virtual reality. World Psychiatry 2021;20:318–35. https://doi.org/10.1002/wps.20883.

[34] Liu X, Liu X, Sun J, Yu NX, Sun B, Li Q, et al. Proactive suicide prevention online (PSPO): machine identification and crisis management for Chinese social media users with suicidal thoughts and behaviors. J Med Internet Res 2019;21, e11705. https://doi.org/10.2196/11705.

[35] Barak-Corren Y, Castro VM, Javitt S, Hoffnagle AG, Dai Y, Perlis RH, et al. Predicting suicidal behavior from longitudinal electronic health records. Am J Psychiatry 2017;174:154–62. https://doi.org/10.1176/appi.ajp.2016.16010077.

[36] Rezaii N, Wolff P, Price BH. Natural language processing in psychiatry: the promises and perils of a transformative approach. Br J Psychiatry 2022;220:251–3. https://doi.org/10.1192/bjp.2021.188.

[37] Fitzpatrick KK, Darcy A, Vierhile M. Delivering cognitive behavior therapy to young adults with symptoms of depression and anxiety using a fully automated conversational agent (Woebot): a randomized controlled trial. JMIR Ment Health 2017;4, e19. https://doi.org/10.2196/mental.7785.

[38] Wechsler TF, Kümpers F, Mühlberger A. Inferiority or even superiority of virtual reality exposure therapy in phobias?—a systematic review and quantitative meta-analysis on randomized controlled trials specifically comparing the efficacy of virtual reality exposure to gold standard in vivo exposure in agoraphobia, specific phobia, and social phobia. Front Psychol 2019;10:1758. https://doi.org/10.3389/fpsyg.2019.01758.

[39] Lundin RM, Menkes DB. Realising the potential of digital psychiatry. Lancet Psychiatry 2021;8:655. https://doi.org/10.1016/S2215-0366(21)00165-6.

[40] Siontis KC, Noseworthy PA, Attia ZI, Friedman PA. Artificial intelligence-enhanced electrocardiography in cardiovascular disease management. Nat Rev Cardiol 2021;18:465–78. https://doi.org/10.1038/s41569-020-00503-2.

[41] Perez MV, Mahaffey KW, Hedlin H, Rumsfeld JS, Garcia A, Ferris T, et al. Large-scale assessment of a smartwatch to identify atrial fibrillation. N Engl J Med 2019;381:1909–17. https://doi.org/10.1056/NEJMoa1901183.

[42] Attia ZI, Noseworthy PA, Lopez-Jimenez F, Asirvatham SJ, Deshmukh AJ, Gersh BJ, et al. An artificial intelligence-enabled ECG algorithm for the identification of patients with atrial fibrillation during sinus rhythm: a retrospective analysis of outcome prediction. Lancet 2019;394:861–7. https://doi.org/10.1016/S0140-6736(19)31721-0.

[43] Shafrin J, Sullivan J, Goldman DP, Gill TM, Ginsberg SD. The association between observed mobility and quality of life in the near elderly. PLoS ONE 2017;12, e0182920. https://doi.org/10.1371/journal.pone.0182920.

[44] Alexander NB, Goldberg A. Gait disorders: search for multiple causes. Cleve Clin J Med 2005;72:586. https://doi.org/10.3949/ccjm.72.7.586.

[45] Colombo G, Joerg M, Schreier R, Dietz V. Treadmill training of paraplegic patients using a robotic orthosis. J Rehabil Res Dev 2000;37:693–700.

[46] Calabrò RS, Cacciola A, Bertè F, Manuli A, Leo A, Bramanti A, et al. Robotic gait rehabilitation and substitution devices in neurological disorders: where are we now? Neurol Sci 2016;37:503–14. https://doi.org/10.1007/s10072-016-2474-4.

[47] Masiero S, Poli P, Rosati G, Zanotto D, Iosa M, Paolucci S, et al. The value of robotic systems in stroke rehabilitation. Expert Rev Med Devices 2014;11:187–98. https://doi.org/10.1586/17434440.2014.882766.

[48] Esquenazi A, Talaty M. Robotics for lower limb rehabilitation. Phys Med Rehabil Clin N Am 2019;30:385–97. https://doi.org/10.1016/j.pmr.2018.12.012.

[49] Nam Y-G, Lee JW, Park JW, Lee HJ, Nam KY, Park JH, et al. Effects of electromechanical exoskeleton-assisted gait training on walking ability of stroke patients: a randomized controlled trial. Arch Phys Med Rehabil 2019;100:26–31. https://doi.org/10.1016/j.apmr.2018.06.020.

[50] Esquenazi A, Talaty M, Packel A, Saulino M. The ReWalk powered exoskeleton to restore ambulatory function to individuals with thoracic-level motor-complete spinal cord injury. Am J Phys Med Rehabil 2012;91:911–21. https://doi.org/10.1097/PHM.0b013e318269d9a3.

[51] Carvalho I, Pinto SM, das Virgens Chagas D, Praxedes dos Santos JL, de Sousa Oliveira T, Batista LA. Robotic gait training for individuals with cerebral palsy: a systematic review and meta-analysis. Arch Phys Med Rehabil 2017;98:2332–44. https://doi.org/10.1016/j.apmr.2017.06.018.

[52] Zeng D, Qu C, Ma T, Qu S, Yin P, Zhao N, et al. Research on a gait detection system and recognition algorithm for lower limb exoskeleton robot. J Braz Soc Mech Sci Eng 2021;43:298. https://doi.org/10.1007/s40430-021-03016-2.

[53] Stroke Foundation. Clinical Guidelines for Stroke Management 2017; 2019.

[54] Qian Z, Bi Z. Recent development of rehabilitation robots. Adv Mech Eng 2015;7, 563062. https://doi.org/10.1155/2014/563062.

[55] La Rovere MT, Pinna GD, Maestri R, Mortara A, Capomolla S, Febo O, et al. Short-term heart rate variability strongly predicts sudden cardiac death in chronic heart failure patients. Circulation 2003;107:565–70. https://doi.org/10.1161/01.CIR.0000047275.25795.17.

[56] Sanderson JE. Heart rate variability in heart failure. Heart Fail Rev 1998;2: 235–44. https://doi.org/10.1023/A:1009745814816.

[57] Beam AL, Kohane IS. Translating artificial intelligence into clinical care. JAMA 2016;316:2368. https://doi.org/10.1001/jama.2016.17217.

[58] Lam TYT, Cheung MFK, Munro YL, Lim KM, Shung D, Sung JJY. Randomized controlled trials of artificial intelligence in clinical practice: systematic review. J Med Internet Res 2022;24, e37188. https://doi.org/10.2196/37188.

[59] Sendak M, Gao M, Nichols M, Lin A, Balu S. Machine learning in health care: a critical appraisal of challenges and opportunities. EGEMS (Wash DC) 2019;7:1. https://doi.org/10.5334/egems.287.

[60] Yu K-H, Beam AL, Kohane IS. Artificial intelligence in healthcare. Nat Biomed Eng 2018;2:719–31. https://doi.org/10.1038/s41551-018-0305-z.

[61] Briganti G, Le Moine O. Artificial intelligence in medicine: today and tomorrow. Front Med 2020;7:27. https://doi.org/10.3389/fmed.2020.00027.

[62] Maddox TM, Rumsfeld JS, Payne PRO. Questions for artificial intelligence in health care. JAMA 2019;321:31. https://doi.org/10.1001/jama.2018.18932.

[63] Quinn TP, Jacobs S, Senadeera M, Le V, Coghlan S. The three ghosts of medical AI: can the black-box present deliver? Artif Intell Med 2022;124, 102158. https://doi.org/10.1016/j.artmed.2021.102158.

[64] Plana D, Shung DL, Grimshaw AA, Saraf A, Sung JJY, Kann BH. Randomized clinical trials of machine learning interventions in health care: a systematic review. JAMA Netw Open 2022;5, e2233946. https://doi.org/10.1001/jamanetworkopen.2022.33946.

[65] Kueper JK, Terry AL, Zwarenstein M, Lizotte DJ. Artificial intelligence and primary care research: a scoping review. Ann Fam Med 2020;18:250–8. https://doi.org/10.1370/afm.2518.

[66] Park Y, Jackson GP, Foreman MA, Gruen D, Hu J, Das AK. Evaluating artificial intelligence in medicine: phases of clinical research. JAMIA Open 2020;3:326–31. https://doi.org/10.1093/jamiaopen/ooaa033.

[67] Buch VH, Ahmed I, Maruthappu M. Artificial intelligence in medicine: current trends and future possibilities. Br J Gen Pract 2018;68:143–4. https://doi.org/10.3399/bjgp18X695213.

[68] Goh KH, Wang L, Yeow AYK, Poh H, Li K, Yeow JJL, et al. Artificial intelligence in sepsis early prediction and diagnosis using unstructured data in healthcare. Nat Commun 2021;12:711. https://doi.org/10.1038/s41467-021-20910-4.

[69] Shaw J, Rudzicz F, Jamieson T, Goldfarb A. Artificial intelligence and the implementation challenge. J Med Internet Res 2019;21, e13659. https://doi.org/10.2196/13659.

[70] Zhang C, Lu Y. Study on artificial intelligence: the state of the art and future prospects. J Ind Inf Integr 2021;23, 100224. https://doi.org/10.1016/j.jii.2021.100224.

[71] Wang X, Han Y, Wang C, Zhao Q, Chen X, Chen M. In-edge AI: intelligentizing mobile edge computing, caching and communication by federated learning. IEEE Netw 2019;33:156–65. https://doi.org/10.1109/MNET.2019.1800286.

[72] You X, Zhang C, Tan X, Jin S, Wu H. AI for 5G: research directions and paradigms. SCIENCE CHINA Inf Sci 2019;62:21301. https://doi.org/10.1007/s11432-018-9596-5.

CHAPTER 7

Medical AI and tort liability

I. Glenn Cohen[a,b], Andrew Slottje[c], and Sara Gerke[d]

[a]James A. Attwood and Leslie Williams Professor of Law, Harvard Law School, Cambridge, MA, United States
[b]Petrie-Flom Center for Health Law Policy, Biotechnology & Bioethics, Harvard Law School, Cambridge, MA, United States
[c]J.D., Class of 2023, Harvard Law School, Cambridge, MA, United States
[d]Penn State Dickinson Law, Carlisle, PA, United States

Introduction

This chapter examines the role of liability in shaping the use of artificial intelligence (AI) in medicine. It begins with a hypothetical vignette, then examines various forms of liability—physician medical malpractice, lack of informed consent, corporate liability for hospitals, and developer liability. Finally, it turns to the preemption of liability and regulation. While many of the points raised apply across legal systems, this chapter focuses primarily on US law, with some reference to EU law.

A starting vignette

Imagine an oncologist providing cancer treatment to a patient in a hospital somewhere in the United States. Imagine that all else equal, the physician in question would administer paclitaxel to the patient at an established dose of $175\,mg/m^2$ body surface area. Now, imagine instead that our physician has access to an AI system developed by the hospital with a software company. Recently cleared by the Food and Drug Administration (FDA), the system recommends a dose that varies over time, based in part on the patient's changing biomarker levels, as measured by hospital employees. While the hospital (and indeed the insurer's guidelines) promotes the use of the AI system, there is no requirement to use it. Indeed, even when using it, the final decision on the correct dosage remains the physician's to make. The physician ultimately chooses to use the medical AI system in setting the dose for this patient. The AI system recommends a dose higher than the established standard of care, and the physician follows the recommendation. The patient experiences side effects that we have reason to believe are the result of the elevated dose, causing physical, emotional, and financial injury [1].

Under the current law, against whom, if anyone, may the patient successfully bring a lawsuit? The physician, the hospital, or the AI developer?

Artificial Intelligence in Medicine
https://doi.org/10.1016/B978-0-323-95068-8.00007-8

If we were system designers able to freely alter the prevailing rules to govern this kind of vignette, what liability rules would be optimal?

While this vignette is hypothetical, AI systems like it have already entered clinical use [2]. Many of these systems use machine learning algorithms that model relationships in data to make predictions. Such systems can help interpret X-rays [3] or determine insulin dosage [4]. Nowadays, these systems are usually "locked" by their developers, but they may also be "adaptive," adjusting to data in the field [5]. Many apply algorithms like neural networks—so-called "deep learning," a subset of machine learning. Such algorithms are considered "black boxes" because they are exceedingly difficult for humans to understand [6]. Thus far, the vast majority of AI systems are intended to support rather than fully determine a clinical decision, though some like Digital Diagnostics's LumineticsCore (formerly called "IDx-DR") are meant to be "autonomous" in the sense of providing, for instance, a complete analysis to a primary care physician as to whether to recommend referral to an ophthalmologist or rescreening in 12 months [7].

The major reason for adopting AI in medicine is its potential to produce better results than those a physician would reach without its assistance. However, demonstrating that a particular AI will achieve this is not easy. It is one thing to show success in development and testing, but quite another to prove that this improvement will be achieved reliably in the real world across different health-care systems—especially because so much of AI's performance may depend on the larger context in which it is embedded [8]. For example, how do physicians react to AI recommendations? Do they over- or undercorrect for what they perceive as its shortcomings? What role is played by insurer decisions that limit reimbursement to AI-recommended treatments?

But even when medical AI demonstrably improves outcomes *overall for a patient population*, as to a particular patient, it may produce a worse outcome than if it had not been used. In a particular case, algorithmic predictions may be inaccurate when based on erroneous data, biased data, or data otherwise poorly matched to the clinical application, when a model's specifications make it too specific or too general, or when a system is administered incorrectly. And the picture is further complicated by system integrity issues, like adversarial attacks and data breaches [9], that are beyond the scope of this chapter.

When such adverse events occur, health-care professionals, health systems, and product developers can face liability claims by the patient who is injured. Understanding the scope of such liability is crucial because it will shape the design and adoption of AI in health care.

Liability for medical AI

A patient injured by medical AI may sue responsible parties in court to recover damages. Liability in such cases is determined according to the principles of tort law. Although case law directly addressing medical AI liability is missing, medical professionals, health systems, and medical device manufacturers may all be held liable when applying general tort law principles. Lawsuits, contracts, and regulations may affect how they share this liability.

Regulation, enforced by administrative liability, may also govern AI design and use. For instance, in the European Union, explainability obligations under the General Data Protection Regulation may inform the use of black-box algorithms in medical AI systems [10]. AI development is further shaped by the US and EU privacy laws.

Tort liability for medical AI: An overview

Medical professionals may be held liable for negligent treatment. In its most basic form, such a claim requires demonstrating that a health professional breached a duty that caused an injury. In medical malpractice, such a breach occurs if a physician provides medical care that deviates from the standard of care. These principles will apply when a physician uses an AI system. Thus, a physician can be liable for injuries from the use of AI technology in diagnosis or treatment when deviating from the standard of care.

Health-care professionals may also be liable for failing to obtain a patient's informed consent for treatment. Again, to simplify, if a (reasonably prudent) patient would have chosen a different treatment with proper disclosure, a physician may be liable for injuries from the treatment [11]. Could the failure to disclose that AI was involved in a physician's recommended course of action lead to liability? As far as we are aware, there have been no reported cases directly on the topic, but under current case law established on informed consent more generally, liability would be most likely when AI decision-making is treated as determinative rather than as an input into the physician's decision, and when the physician's control over a procedure, or level of experience with that procedure, seems relevant to the patient's choice. One example of possible informed consent liability might be that of a physician who fails to disclose AI use in AI-driven surgery.

Health systems, physician groups, and physician-employers may also be held vicariously liable for professional negligence. Employers, as principals, may be vicariously liable for the negligence of their employees in some instances while vicarious liability for independent contractors is possible, but more challenging, to establish. This distinction according to employment status matters, especially in the United States, where nurses are often employees while physicians are more traditionally independent contractors. A court might still impute vicarious liability to a health system for a physician's use of AI on theories such as apparent agency or agency by estoppel—for example, when an injured patient has relied on representations of safety made by a hospital advertising its provision of AI-driven medicine [12].

Health systems can also be directly liable for injuries. A health system has its own traditional duties of care to patients such as providing safe facilities and equipment. The emergence of corporate negligence doctrine has enabled some patients to bring claims for negligence in supervising medical care or in enforcing hospital rules about patient care [13]. For example, a negligent credentialing claim alleges that a hospital should have known a physician at its premises was not qualified to practice. It is possible that we will see theories of corporate negligence brought to courts in relation to AI adoption. Negligent credentialing or supervision theories might be extended to treat medical AI like an additional physician: to put it pithily, we could think of hospital systems as "hiring" not "buying" an AI [14]. A health system might thus be liable for making AI technology available to practitioners without proper vetting or validation. As with medical providers, the direct liability of hospital systems will evolve as standards for AI use change [15].

Turning to the development of AI, manufacturers can be liable for injuries from product defects. Liability for selling a defective product is strict, with no need to show negligence [16]. While the risks intrinsic to medicine need not render a medical product defective, sellers can be liable for manufacturing defects, errors in product design, and failures to provide adequate warnings. Showing a design defect may entail showing that a product is unreasonably dangerous under the "risk–utility" test, which may require showing that the product could have feasibly been designed in a better way.

To an extent, current case law may inform such liability. Some types of defects in medical AI, like mechanical flaws in a product's hardware, resemble traditional sources of product liability. Health systems and medical professionals traditionally do not face liability for product defects as sellers [17], so strict liability claims may be more likely against developers. On the other hand, a health system may itself participate in product development—for example, by sharing data with a software developer.

Claims about defective AI software may also face several particular barriers [18]. It is not clear that software is a product, as opposed to a service [19]. Increased system autonomy might impede attribution of responsibility to the product's manufacturer. Because AI can have multiple developers, it may be unclear who the manufacturer is in any case. Although errors in dataset choice or algorithmic specification are plausibly design defects, courts, mindful of the needs of innovation, seem generally hesitant to review medical product designs. Algorithmic opacity may make it difficult to perform the cost-benefit analysis necessary under the risk-utility test. And the learned intermediary doctrine limits failure-to-warn liability for medical products on the premise of the physician's intervening judgment, a rationale that may come into question if AI distances physicians from treatment decisions.

Apportioning liability

Now that we have a basic sense of the overall medical AI liability landscape, we can examine how the various liable parties may interact to reapportion that liability. A tort plaintiff may recover from multiple responsible parties. Historically, manufacturers and health-care providers might have shared fully in responsibility for an "indivisible injury" [20], putting the burden on defendants to apportion damages in actions for contribution. Changes in US tort law have increased up-front apportionment [21]. Thus, if the patient in the vignette sues the treating physician, the hospital, and the AI developer, tort liability could be apportioned among the parties according to their level of fault as determined in court.

Some of this liability may be shifted by contract. For example, a health system may contract to indemnify a practitioner's malpractice liability, or a developer may put hold harmless or arbitration clauses in a contract with a health system [22]. Practitioners and health systems may purchase insurance, shifting liability risk to insurers. Indemnity and insurance could expand the medical AI market by diminishing physician liability, by clarifying its uncertainty, or by establishing rules or incentives for the use of medical AI [23]. Developers could also indemnify the health systems that purchase and implement their products—the LumineticsCore has apparently done exactly this [24], though its eagerness to do so may also stem from the fact that it was first-in-class and needed to mollify more concerns to build the market than will successor products.

Patients are less likely to contract away their rights to sue. Insurers have viewed arbitration agreements as economically unfavorable, and some courts have taken a dim view of contracts that compel injured patients to arbitrate [25]. But after an injury, health systems may offer payment in exchange for a patient agreeing not to sue [26]. Uncertainty about liability for medical AI use could increase the risks of litigation, affecting the incentives for health systems to enable, and for patients to enter, such pre-litigation settlement agreements.

Regulation may preempt product liability or malpractice claims, shifting a controlled level of risk onto the public. Programs like the FDA approvals for medical devices may substitute ex ante regulatory requirements for ex post liability. Administrative remedies can shift a controlled level of liability onto the government or mandatory insurance programs. We discuss regulatory preemption further below.

There are interesting divergences between how tort law assigns responsibility and the perspectives of the public on who is responsible. One study indicates that the public might generally impute greater fault for medical AI errors to medical professionals and health systems than developers [27]. Previous decisions in other settings have similarly conveyed hesitance to excuse physician errors in administering defective products [28]. The relatively heavier share of liability that juries might apportion to physicians and health systems could make malpractice liability particularly important in shaping the adoption of medical AI, and we turn to it next.

Malpractice liability and the standard of care

In this section, we take a deeper dive into what malpractice actions against physicians and other medical professionals may look like when AI is involved. Malpractice actions require the breach of a duty of care, but jury members may have little experience or even intuition of the relevant baseline of appropriate care [29]. Some jurisdictions set a standard of care that reflects the customary practice for physicians with a similar level of training and resources. (Some even follow the historical practice of setting this standard with reference to local professionals.) Some jurisdictions instead impose a standard of "reasonable care," ideally allowing liability to track the state of the art instead of custom [30]. Standards of care may also be set by statute. Applying the standard of care to technological advances may prove tricky: an established standard is at odds with the use of new technologies, but reasonable care may demand it.

When using AI deviates from the standard of care, the physician risks malpractice liability for resulting injury. By contrast, if the physician adheres to the standard of care—whether it benefits the patient or not—malpractice liability is unlikely. These principles and their implications are played out in Table 7.1 [31]. The punchline is that the current set-up of medical

Table 7.1 Application of malpractice to medical AI.

AI recommendation consistency with standard of care	Correctness of AI recommendation	Physician action	Outcome	Legal liability
Consistent with standard of care	Correct	Follows AI	No injury	No liability
		Does not follow AI	Injury	Liability (did not follow standard of care)
	Incorrect	Follows AI	Injury	No liability (followed standard of care)
		Does not follow AI	No injury	No liability
Inconsistent with standard of care	Correct	Follows AI	No injury	No liability
		Does not follow AI	Injury	No liability (followed standard of care)
	Incorrect	Follows AI	Injury	Liability (did not follow standard of care)
		Does not follow AI	No injury	No liability

malpractice liability provides an incentive *not* to use AI when it is contrary to the ordinary course of treatment a physician would undertake. To return to our starting vignette, this would mean a physician is incented to stick with the standard dose of the chemotherapeutic agent instead of the more patient-specific dose recommended by the AI system. When that helps avoid an adverse event, that seems like a good result. But even when the AI produces better outcomes at the population level, and the dose it recommends would be safer and more effective for the patient, the physician is still incented to ignore it. Put otherwise, the chief value of the medical AI is when it tells us to do something *other* than what the physician would otherwise do, but medical malpractice liability will deter physicians from departing from the standard treatment in precisely such cases.

This unfortunate result is not inevitable. Incorporating the AI system's recommendation into the standard of care can enhance the incentives to use AI. If the AI recommendation is incorporated as a recognized alternative treatment under the "two schools of thought" doctrine, there is limited malpractice liability either for following the AI recommendation or for following the pre-AI standard of care. Litigation risk then provides an incentive to choose the alternative least likely to cause injury, which may be the AI system. A more extreme possibility would be if the AI recommendation becomes a required part of the standard of care. Then it is failing to follow the AI recommendation that creates liability risk by deviating from the standard of care. In that case, malpractice liability provides an incentive to use AI.

How do we get to this better place for standard of care? One possibility is that juror impressions might get us there. One study of mock jurors found that members of the public were likely to view a physician's decision to follow a nonstandard AI recommendation that later turned out to be incorrect as reasonable [32]. On the other hand, the mock jurors were more indecisive about whether a physician's decision to deviate from a correct nonstandard AI recommendation was acceptable. Such views suggest an emerging perception of AI as something like a recognized alternative treatment. But such changes in perception may require time and may be "lumpy" in that they may depend heavily on the way evidence is framed in particular trials as well as orthogonal elements of the case (such as the severity of the injury that resulted). Otherwise put, this mechanism of shifting the standard of care may not be certain enough to quell the concerns of physicians in following AI recommendations that deviate from what they would otherwise do.

Another way the standard of care might evolve is through medical practice guidelines. These guidelines have become increasingly important in malpractice cases as deference to medical customs has eroded. Practice guidelines from professional medical societies like the American Medical Association and specialty boards, health systems, insurers, and federal and state entities like the Agency for Healthcare Research and Quality may all be used as evidence of the standard of care [33]. Guidelines might thus be written to encourage or discourage the use of medical AI in particular contexts depending on its assessed risks and benefits.

The standard of care may vary for different types of AI systems. Some medical software might provide transparent information to support a final physician decision—for example, by reporting patient records or calculating well-understood relationships. This model of physician decision-making resembles the current one, so liability may resemble the current regime. Other AI products transform input data into treatment decisions in a less transparent way. The physician may choose whether to use AI but may lack any basis to assess its recommendations.

Alternatively, some products might fall in between in terms of the transparency of the decision. For example, an algorithm that uses a convolutional neural network architecture for image-based diagnosis might provide a heatmap showing where the algorithm is "looking," giving a physician limited information that may nonetheless fail to explain the decision completely [34].

For more opaque types of AI systems, courts could assess liability with reference to the physician's initial decision to use medical AI. Even when an AI system provides a limited explanation, its recommendations may be most valuable when they most supplement existing decision-making—that is, counterintuitively, when its explanation seems most inscrutable [35]. Liability that leads to second-guessing AI recommendations might close off these especially useful applications [36]. Instead of explainability, courts might look to "indicia of reliability," like regulatory approvals or empirical success in randomized trials [37]. On such indicia, a physician's initial decision to use AI may be reasonably calculated to benefit the patient despite a foreseeable risk that the AI could be wrong. Avoiding negligence liability in such a situation seems to track the question of the standard of care, and avoids imposing strict liability for the AI's errors. Courts seem to have followed such an approach with other products, like pharmaceuticals, where physicians must similarly rely on indicia of reliability like studies and approvals [38].

The standard of care may also come to incorporate other indicia of reliability, like outside validation. For systems where physicians are poorly positioned to second-guess an AI's decisions, validation procedures, like requiring predictions to match third-party computation or requiring different AI systems to agree on a prediction, could help ensure that AI systems are functioning properly [39]. Such procedural standards may be more administrable than requiring the physician to assess an opaque system's recommendations and could be defined through regulation or practice guidelines.

As AI use increases, best practices may emerge that provide evidence of the standard of care. If greater use of AI systems leads patterns to surface in the types of ways that systems can fail, the emergence of predictable forms of harm may establish the foreseeability necessary for courts to discern breach [40].

And apart from breach, the plaintiff's causation and injury burdens may already be challenging to carry in a malpractice case. Algorithmic opacity could make it even more difficult to prove causation, a problem that could create incentives to select less transparent systems [41]. And errors in AI development that moderately impair a model without bankrupting it of predictive force, like dataset bias, could increase the risk of adverse outcomes without causing injury in the traditional doctrinal sense.

Regulation and preemption as alternative to the common law of torts

We have thus far focused on the application to medical AI of the common law of torts that serves as the background principle for liability in the US system. Federal and state regulation can preempt common-law liability in exchange for more predictable regulatory requirements or administrative remedies. FDA regulation is unlikely to preempt malpractice liability for the use of AI in medical practice, but it could preempt some manufacturer liability for product defects. Federal or state legislatures might also preempt malpractice or product liability by establishing specialized adjudication procedures, mandatory insurance schemes, or general compensation funds for AI-related medical injuries. Regulation that diminishes or clarifies liability for medical AI might encourage its adoption.

FDA regulation

The Federal Food, Drug, and Cosmetic Act (FDCA) makes the FDA responsible for regulating medical devices. Broadly speaking, devices are intended to diagnose, cure, mitigate, prevent, or treat diseases [42]. The 21st Century Cures Act excludes from this definition some software functions that provide health-care professionals recommendations to consider in making independent decisions, as opposed to software functions that provide the primary basis for decision-making [43]. The AI products outside this exemption that are intended for use in diagnosis, cure, mitigation, prevention, or medical treatment are likely subject to FDA regulation as medical devices, and AI systems could also be regulated as accessories to regulable devices [44].

FDA device regulation preempts some tort liability. In general, the 1976 Medical Device Amendments (MDA) to the FDCA displace state law concerning medical devices for human use [45], which the Supreme Court has interpreted to preclude some state-law tort claims. In *Medtronic v. Lohr*, the

Court considered a device exempt from the rigorous premarket approval (PMA) process because of its substantial equivalence to a pre-1976 device [46]. The Court found that design defect claims were not preempted because the FDA had reviewed the device under the 510(k) process for only equivalence, not product safety, and that federal manufacturing and labeling requirements were too general to preempt related state-law claims [47]. Subsequently, in *Riegel v. Medtronic*, the Court, considering a device that went through PMA [48], applied a two-part test. First, the Court considered "whether the Federal Government ha[d] established requirements applicable to" the device that could preempt state law. Unlike the 510(k) substantial equivalence process in *Lohr*, the PMA process imposed such requirements [49]. Second, the Court considered whether the state-law claims were grounded in safety- and effectiveness-related "requirements with respect to the device that are 'different from, or in addition to,' the federal ones," viewing "common-law duties" as imposing such requirements [50].

These decisions could leave developers subject to some liability for defective medical AI. First, outside the tort system, developers are subject to liability under the FDCA for regulatory infractions [51]. Second, express preemption under the MDA applies a priori only to devices for human use. Nondevice software, such as under the 21st Century Cures Act exclusion, is shielded by more limited forms of preemption, if at all [52]. Third, most AI-based medical devices have entered the market via the 510(k) process [53]. Thus, manufacturers are not shielded from state-law tort claims challenging the safety and effectiveness of such devices. Substantial equivalence, as in *Lohr*, does not impose "requirements" at the first step of *Riegel*. And fourth, *Riegel* noted that state courts could still provide causes of action that paralleled the violation of FDA requirements rather than being "different from, or in addition to" those requirements [54].

One open question in this area has been whether parallel state-law claims may be grounded in industry-wide federal standards, like the FDA's Quality System Regulations or Current Good Manufacturing Practices (CGMPs). The U.S. Court of Appeals for the Eighth Circuit answered this question in the negative, citing the generic nature of such requirements [55]. (The district court had pointed to the fact that the FDA required "manufacturers to develop *their own* quality-system controls" [56].) But rejecting this functionalist analysis, the Seventh Circuit subsequently noted that the language of the preemption provisions did not distinguish between "general" and "device-specific" requirements [57].

Another salient question is the degree to which claims are preempted under FDCA Section 310(a), which limits enforcement of the FDCA to the federal government [58]. Appellate courts have split on whether some state-law claims grounded in FDCA violations should be viewed as "attempt[s] by private parties to enforce" FDCA requirements [59] or as claims for "breach of a recognized state-law duty" [60].

Emerging requirements for software developers seem to tread near these fault lines, indicating the potential for disputes about whether FDCA requirements preempt liability for defective medical AI. For instance, proposed developer precertification standards may delegate requirements for "quality and organizational excellence" to industry [61], similarly to the CGMPs. That may invite litigation as to whether those standards are too generic to support parallel state-law claims. Similarly, reliance on post-market monitoring could lead to state-law failure-to-warn claims grounded in violations of federal reporting requirements, but the implied preemption of such claims under FDCA Section 310(a) may be in dispute [62]. At a general level, if the need to accommodate the quicker innovation pace or adaptive capabilities of AI systems leads to the backgrounding of premarket review and a heavier emphasis on postmarket monitoring, the rationale that ex ante regulation by the FDA substitutes for ex post regulation by tort may come into question, leading courts to interpret preemption narrowly.

Similarly, the 21st Century Cares Act's exclusion of some software functions from the medical device definition under the FDCA removes the shield of express preemption. That has the effect of shifting regulatory responsibility for such software onto the tort system. Yet the exclusion only applies to some software functions that are transparent clinical decision support tools [63]. The medical device exception might thus enable ex post liability for software functions with more ex ante transparent effects while leaving the systems whose effects are more challenging to ascertain ex ante potentially shielded from liability ex post provided they will be approved via PMA (rather than 510(k)).

Remedies outside of tort law

Governments could also impose administrative programs that preempt liability for the use of medical AI while providing nontort remedies. One such model is the National Vaccine Injury Compensation Program, which has vaccine manufacturers pay into a system that limits common-law tort claims and provides an administrative forum for claims instead [64]. Likewise, Florida and Virginia programs provide no-fault compensation for some birth-related injuries [65]. These programs were intended to increase the provision of covered services by preempting tort liability. But if such programs are not designed carefully and imposed by clear statutes, courts can construe the preemption narrowly [66], sometimes putting defendants in the unhappy position of both funding the program *and* being called to account in court under ordinary tort law.

If designed carefully, administrative adjudication and no-fault compensation programs might shape incentives for AI innovation and adoption. The special difficulties in showing causation of injury where opaque AI systems are involved could make administrative adjudication or compensation funds more responsive fora in some respects. Accordingly, the European Parliament has suggested considering compulsory insurance or a general compensation fund to address "the complexity of allocating responsibility" when autonomous systems inflict injury [67].

Conclusion

In this chapter, we have tried to demonstrate that the question of how traditional tort law will accommodate the use of medical AI is complex and very much evolving. There are multiple forms of liability that apply differently to different players in the ecosystem (physicians, hospitals, and developers). Many decisions about what to develop and what to adopt are driven by expectations about what tort law will do. Through contractual apportionment of liability and indemnification, the players have some freedom to redesign the liability terms, but it may not be enough to harness the full value of medical AI. Preempting some forms of tort liability and adopting alternative compensation schemes through administrative processes may prove more attractive in the long run.

Acknowledgments

I. Glenn Cohen's work on this chapter was supported by a Novo Nordisk Foundation Grant for a scientifically independent International Collaborative Bioscience Innovation & Law Programme (Inter-CeBIL programme - grant no. NNF23SA0087056). Sara Gerke's work was funded by the European Union (Grant Agreement no. 101057321). Views and opinions expressed are however those of the author(s) only and do not necessarily reflect those of the European Union or the Health and Digital Executive Agency. Neither the European Union nor the granting authority can be held responsible for them. S.G. also reports grants from the European Union (Grant Agreement no. 101057099), the National Institute of Biomedical Imaging and Bioengineering (NIBIB) and the National Institutes of Health Office of the Director (NIH OD) (Grant Agreement no. 3R01EB027650-03S1 and no. 1R21EB035474-01), and the National Institute on Drug Abuse (NIDA) and the National Institutes of Health (NIH) (Grant Agreement no. 1U54DA058271-01).

References

[1] Maier C, Hartung N, Kloft C, Huisinga W, de Wiljes J. Reinforcement learning and Bayesian data assimilation for model-informed precision dosing in oncology. CPT Pharmacometrics Syst Pharmacol 2021;10:241.

[2] Maliha G, Gerke S, Cohen IG, Parikh R. Artificial intelligence and liability in medicine. Milbank Q 2021;99:629–34.

[3] Gerke S, Minssen T, Cohen IG. Ethical and legal challenges of artificial intelligence-driven healthcare. In: Bohr A, Memarzadeh K, editors. Artificial intelligence in healthcare; 2020. p. 299.

[4] Benjamens S, Dhunnoo P, Meskó B. The state of artificial intelligence-based FDA-approved medical devices and algorithms. NPJ Digit Med 2020;3. https://www.nature.com/articles/s41746-020-00324-0.

[5] Babic B, Gerke S, Evgeniou T, Cohen IG. Algorithms on regulatory lockdown in medicine. Science 2019;366:1202.

[6] Babic B, Gerke S, Evgeniou T, Cohen IG. Beware explanations from AI in health care. Science 2021;373:284. Babic B, Gerke S. Explaining medical AI is easier said than done. STAT 2021;(July 21). https://www.statnews.com/2021/07/21/explainable-medical-ai-easier-said-than-done/.

[7] Gerke S, Babic B, Evgeniou T, Cohen IG. The need for a system view to regulate artificial intelligence/machine learning-based software as medical device. NPJ Digit Med 2020;3. https://www.nature.com/articles/s41746-020-0262-2. IDx-DR, Digit Diagnostics, https://www.digitaldiagnostics.com/products/eye-disease/idx-dr/ (last visited Feb. 26, 2022).

[8] See Gerke et al., *supra* note 7.

[9] Finlayson SG, Bowers JD, Ito J, Zittrain JL, Beam AL, Kohane IS. Adversarial attacks on medical machine learning. Science 2019;363:1287.

[10] Note that the existence and content of a right to explanation under the General Data Protection Regulation is contested. *See, e.g.*, Wachter S, Mittelstadt B, Floridi L. Why a right to explanation of automated decision-making does not exist in the general data protection regulation. Int Data Privacy Law 2017;7:76.

[11] Cohen IG. Informed consent and medical artificial intelligence. Georgetown Law J 2020;108:1425–32.

[12] See Price WN II, S. Gerke, I.G. Cohen. Liability for use of artificial intelligence in medicine. In: B. Solaiman, I.G. Cohen, editors. Research handbook in health, AI and the law; forthcoming. Alstott A. Hospital liability for negligence of independent contractor physicians under principles of apparent agency. J Legal Med 2004;25:487–90.

[13] Furrow B. Medical malpractice liability: of modest expansions and tightening standards. In: Cohen IG, Hoffman AK, Sage WM, editors. The Oxford Handbook of U.S. Health Law; 2017. p. 421–34.

[14] Gerke et al., *supra* note 7.

[15] Price WN II. Medical malpractice and black-box medicine. In: Cohen IG, Lynch HF, Vayena E, Gasser U, editors. Big data, health law, and bioethics; 2018. p. 295–303.

[16] Manufacturers may also be held liable for product defects on other theories, such as negligence or breach of warranty.

[17] Hollander v. Sandoz Pharm. Corp., 289 F.3d 1193, 1217 n.22 (10th Cir. 2002).

[18] Schweikart SJ. Who will be liable for medical malpractice in the future? How the use of artificial intelligence in medicine will shape medical tort law. Minn J Law Sci Technol 2021;22:1. Selbst AD. Negligence and AI's human users. Boston Univ Law Rev 2020;100:1315.

[19] Evans BJ, Pasquale F. Product liability suits for FDA-regulated AI/ML software. In: Cohen IG, Minssen T, Price WN II, Robertson CT, Shachar C, editors. The future of medical device regulation: Innovation and protection. Cambridge University Press; 2022.

[20] *See, e.g.*, Temple v. Synthes Corp., 498 U.S. 5, 7 (1990) (per curiam).

[21] Restatement (third) of torts: Apportionment of liability. § E18 cmt. b (2000).

[22] Koppel R, Kreda D. Commentary, health care information technology vendors' "hold harmless" clause. JAMA 2009;301:1276.

[23] *See* Ben-Shahar O, Logue KD. Outsourcing regulation: how insurance reduces moral hazard. Mich Law Rev 2012;111:197–203.

[24] Panel: Understanding algorithms, artificial intelligence, and predictive analytics through real world applications. Federal Trade Commission; November 13, 2018 [comments of Michael D. Abramoff] https://www.ftc.gov/system/files/documents/public_comments/2018/11/ftc-2018-0101-d-0004-162932.pdf.

[25] Furrow, *supra* note 13, at 431.

[26] *Id.* at 432 & n.79.

[27] Khullar D, Casalino LP, Qian Y, Yuan L, Chang E, Aneja S. Public vs physician views of liability for artificial intelligence in health care. J Am Med Inform Ass'n 2021;28:1574.

[28] *E.g.*, Bush v. Thoratec Corp., 13 F. Supp. 3d 554, 575 (E.D. La. 2014); *see* Maliha et al., *supra* note 2, at 632.

[29] Greenberg M. Medical malpractice and new devices: defining an elusive standard of care. Health Matrix 2009;19:423. Price WN II, Gerke S, Cohen IG. Potential liability for physicians using artificial intelligence. JAMA 2019;322:1765.

[30] *E.g.*, Nowatske v. Osterloh, 543 N.W.2d 265, 272 (Wis. 1996).

[31] Adapted from Price et al., *supra* note 29.

[32] Price WN II, Gerke S, Cohen IG. How much can potential jurors tell us about liability for medical artificial intelligence? J Nucl Med 2021;62:15. Tobia K, Nielsen A, Stremitzer A. When does physician use of AI increase liability? J Nucl Med 2021;62:17.

[33] Mello MM. Of swords and shields: the role of clinical practice guidelines in medical malpractice litigation. Univ Pennsylvania Law Rev 2001;149:645–65. Government agency guidelines may assume particular influence in light of the Affordable Care Act. *See* Furrow, *supra* note 13, at 426–27.

[34] Harned Z, Lungren MP, Rajpurkar P. Machine vision, medical AI, and malpractice. Harv J Law Technol Digest 2019;5. https://jolt.law.harvard.edu/assets/digestImages/PDFs/Harned19-03.pdf.

[35] Tobey D. Explainability: where AI and liability meet. DLA Piper 2019;(Feb. 25). https://www.dlapiper.com/en/us/insights/publications/2019/02/explainability-where-ai-and-liability-meet/.

[36] On the other hand, if a system's explanation of its decision supports a physician's choice, the physician might seek to adduce the explanation as evidence of due care.

[37] *See* Cohen, *supra* note 11, at 1443.

[38] *See id.*; Richardson v. Miller, 44 S.W.3d 1, 16–17 & nn.19–20 (Tenn. Ct. App. 2000).

[39] Price WN II. Black-box medicine. Harv J Law Technol 2015;28:419–41.

[40] Karnow CEA. The application of traditional tort theory to embodied machine intelligence. In: Calo R, Froomkin AM, Kerr I, editors. Robot law; 2016. p. 51–76.

[41] McNair D, Price WN II. Health care AI: law, regulation, and policy. In: Matheny M, Israni ST, Ahmed M, Whicher D, editors. Artificial intelligence in health care. National Academy of Sciences; 2019. p. 197–233.

[42] FDCA Section 201(h).

[43] FDCA Section 520(o)(1)(E).

[44] Price WN II. Regulating black-box medicine. Mich Law Rev 2017;116:439–40.

[45] FDCA Section 521(a).

[46] *Medtronic v. Lohr*, 518 U.S. 470, 493–94 (1996).

[47] *Id.* at 493–94, 501.

[48] *Riegel v. Medtronic*, Inc., 552 U.S. 312, 320 (2008).

[49] *Id.* at 321, 323.

[50] *Id.* at 321–24; *see also* FDCA Section 521(a).

[51] FDCA Section 303(f).

[52] Developers of such software could argue based on the 21st Century Cures Act that tort liability would "frustrate the achievement of congressional objectives." *Wyeth v. Levine*, 555 U.S. 555, 581 (2009). Where it does have authority to regulate, FDA might also preempt state law tort claims by adopting regulations in direct conflict with such claims. *Id.* at 582 (Breyer, J., concurring).

[53] Artificial intelligence and machine learning (AI/ML)-enabled medical devices. FDA; December 6, 2023. https://www.fda.gov/medical-devices/software-medical-device-samd/artificial-intelligence-and-machine-learning-aiml-enabled-medical-devices.

[54] *Riegel*, 552 U.S. at 330. *See generally* Tarloff ES. Note, *medical devices and preemption*. N.Y.U. Law Rev 2011;86:1196.

[55] In re Medtronic, Inc., Sprint Fidelis Leads Prod. Liab. Litig., 623 F.3d 1200, 1206 (8th Cir. 2010).

[56] In re Medtronic, Inc. Sprint Fidelis Leads Prod. Liab. Litig., 592 F. Supp. 2d 1147, 1157 (D. Minn. 2009) (emphasis in original).

[57] Bausch v. Stryker Corp., 630 F.3d 546, 555 (7th Cir. 2010).

[58] Buckman Co. v. Plaintiffs' Legal Comm., 531 U.S. 341, 353 (2001).

[59] *In re Medtronic*, 623 F.3d at 1205.

[60] *Bausch*, 630 F.3d at 557–58; *accord* Hughes v. Bos. Sci. Corp., 631 F.3d 762, 775 (5th Cir. 2011); Howard v. Zimmer, Inc., 718 F.3d 1209, 1210 (10th Cir. 2013).

[61] Proposed regulatory framework for modifications to artificial intelligence/machine learning (AI/ML)-based software as a medical device (SaMD). FDA; April 2, 2019. https://www.fda.gov/media/122535/download.

[62] *Compare In re Medtronic*, 623 F.3d at 1205, *with* Hughes, 631 F.3d at 775.

[63] FDCA Section 520(o)(1)(E).

[64] 42 U.S.C. § 300aa–10 *et seq.*; *see Schafer v. Am. Cyanamid Co.*, 20 F.3d 1, 2 (1st Cir. 1994) (Breyer, C.J.).

[65] Fla. Stat. Ann. § 766.303 (West 2021); Va. Code Ann. § 38.2-5002 (West 2021).

[66] *See* Moss v. Merck & Co., 381 F.3d 501, 503–04 (5th Cir. 2004); *Schafer*, 20 F.3d at 7; Engstrom NF. Exit, adversarialism, and the stubborn persistence of tort. J Tort Law 2013;6:75.

[67] European Parliament Resolution of 16 February 2017 with Recommendations to the Commission on Civil Law Rules on Robotics (2015/2103(INL)), Eur. Parl. Doc. P8_TA(2017)0051 (2017), ¶¶ 57, 59, https://www.europarl.europa.eu/doceo/document/TA-8-2017-0051_EN.html.

CHAPTER 8

Regulation of AI in healthcare

Colin Gavaghan
School of Law/Bristol Digital Futures Institute, University of Bristol, England, United Kingdom

Introduction

The role of artificial intelligence (AI) in healthcare has been much anticipated. Its potential benefits in this area are not hard to imagine. Healthcare
is an area where demand seems invariably to outpace supply and complaints
about waiting times for treatment and healthcare rationing are a ubiquitous
feature of modern life in developed nations. Demographic changes are likely
to see demands on these services increase. Insofar as AI—and related technologies such as robotics and data analytics—can play a role in meeting these
demands, its arrival is likely to be welcomed.

Moreover, the vast and ever-increasing quantity of research and data is
raising serious questions about the capacity of human doctors to make optimal decisions for patients. AI offers the prospect of churning tirelessly
through new scientific data and making unflaggingly logical recommendations unaffected by human cognitive limitations.

But healthcare is also an area where the arrival of AI has invoked the most
scrutiny, concern, and even resistance. It is also a profession that is highly
regulated, by statute and case law, and by various codes and guidelines. This
chapter will consider how AI is likely to fit within existing regulatory frameworks around healthcare, the sort of regulatory challenges it could present,
and the potential strategies available to lawmakers and regulators to
address them.

As may be expected, the rules vary between jurisdictions. This chapter
does not attempt to give a practical guide to navigating the complexities of
any particular system, but instead focuses on the types of questions that are
arising in all of them.

At the time of writing, the regulatory landscape around AI in healthcare
is in many countries still a state of flux. In the United States, the Food and
Drug Administration recently completed a consultation process on its proposed new regulatory framework [1], and has now published an "action
plan" setting out the next steps for its process [2]. A similar picture is present

Artificial Intelligence in Medicine
https://doi.org/10.1016/B978-0-323-95068-8.00008-X

105

in the United Kingdom [3], with the additional complication that the exit from the EU has required various transitional measures to be introduced. And in New Zealand, a new statute governing therapeutic devices, that will for the first time explicitly provide for AI and other software, is still working its way through the legislative process [4]. With many of these new approaches still undergoing consultation or in draft form, it's likely to be some time before a clear picture of the regulatory terrain emerges. What we can presently see through is that most of these jurisdictions are grappling with the same sorts of regulatory challenges and choices. This chapter will consider some of these issues.

What do we mean by "regulation"?

"Regulation" is a much-discussed and somewhat contested term in legal literature. Some writers define it quite restrictively, using the term only to describe formal legal rules. Others have somewhat more flexible understandings. According to Susy Frankel and John Yeabsley, "Regulation includes legislation, legal rules, codes of practice (both formal and informal), and a combination of these." [5]. Others would extend the term to include self-regulation and even "unintentional influence such as market forces." [6].

Wherever we draw the line for what counts as "Regulation," one thing we can say for certain is that healthcare has attracted a wide variety of laws, rules, guidelines, codes of practice, and such, and these vary in many ways. When it comes to considering the appropriate rules for AI in healthcare, a similarly broad range of options is available. On one end of the spectrum, there could be relatively "hard-edged" legal rules. Most obviously, these can be in the form of legislation—"the law on the statute books." This might be passed by the relevant parliament, though there can also be delegated legislation introduced by Government Ministers [7]. It could also include court judgments [8].

Legal rules can also be made by regulatory agencies with delegated responsibility. In the realm of assisted reproductive technologies, organizations like the UK's Human Fertilisation and Embryology Authority and New Zealand's Advisory Committee on Assisted Reproductive Technology (ACART) are empowered by statute to set limits and conditions on their use [9]. In a similar fashion, regulators like the US Food and Drug Administration, Singapore's Health Sciences Authority and the UK's Medicines and Healthcare product Regulatory Agency (MHRA) have a range of delegated powers to make decisions about medicines and medical products.

Perhaps confusingly, the rules set by ACART are referred to as "guidelines," but they have the status of "secondary legislation." For the most part, publications calling themselves "guidelines" will lack this legislative status, but may still be legally important. These may include explanations from regulators about how the rules work, or guidance from professional organizations. As Joanna Manning has explained:

> The more prestigious the professional source and the greater their uptake in practice, the more likely it will be that clinical guidelines will be perceived to be authoritative and adherence or non-adherence will be taken as a measure of the standard of care [10].

As Manning explains, guidelines "issued by a regulator, professional registration authority or a professional college" may well be regarded as setting the legal standard of care for practitioners. In an area like medical device regulation, where the legislation is notoriously unwieldy and difficult to navigate, guidelines from regulators can provide a valuable tool for manufacturers and users. Many organizations and groupings issue publications, variously described as guidance, codes of practice, codes of ethics, and principles [11]. These can be useful and informative and could even be introduced as evidence of whether a practitioner's conduct was of an appropriate standard, but they will not be legally binding in themselves.

This chapter will be focusing on regulation in the legally significant sense: legal rules as well as those guidelines issued by professional bodies that are likely to have legal implications.

What should be regulated?

So far, a substantial body of codes, guides, and principles have been published about AI, in general and in healthcare. What we have not yet seen are many legal rules or formal guidelines specifically targeted at AI. Does this mean that AI is completely unregulated?

As technology law scholars Roger Brownsword and Morag Goodwin have said, "it will rarely be true to say that an emerging technology finds itself in a regulatory void" [12]. While it's true that AI-specific regulation is still scarce, AI will sometimes be covered by existing laws and rules. Whether those existing rules are adequate depends on the extent to which AI differs from the previous targets of those rules—medical products and devices, processes, and perhaps healthcare practitioners. Not all new technologies require new rules, but it may be that AI is different in respects that make the old rules a poor fit.

In considering whether AI in healthcare needs new regulatory approaches, regulators will need to consider how AI functions and what risks it poses. Nicolas Terry said "The regulatory determinations must be made on the basis of AI risks and benefits, not the risks and benefits of what they substituted for." [13]. To take one example, rules focused on preventing intimate or financial relations between healthcare providers and patients may be unnecessary in the context where AI takes over some of those roles (at least for now! [14]), but at the same time, new considerations may well arise. For example, as discussed in more detail below, the capacity of some AI to adapt and change once in use may raise questions about the adequacy of existing rules about medical products.

If AI-specific rules are deemed necessary, then an important task will be defining their scope and target. Defining the technology of interest can be a tricky task for regulators, especially when dealing with something like AI. It's widely recognized that there is no universally shared definition of "AI" [15]. This isn't entirely a novel problem for law, which often deals with concepts that have no universally accepted usage. For instance, it often relies on concepts like "reasonableness" or "intent," which are highly disputed even among legal scholars. But over time, a body of case law has developed, offering greater certainty about how such concepts will be understood in a particular legal context. If we are dealing with new rules, though, no such guidance will be available, and we should at least hope that they are as clear as they can be made to be. At the very least, regulators will need to know whether a given product falls within the boundaries of their remit, and manufacturers will need to know what route to follow in importing or bringing their products to market.

On the other hand, overprecise regulatory definitions can pose challenges too. Regulators dealing with emerging technologies often face the challenge of what Brownsword has called "descriptive disconnection." This describes the situation when "the covering descriptions employed by the regulation no longer correspond to the technology or to the various technology-related practices that are intended targets for the regulation." [16]. Too precise a regulatory definition of a concept like "AI" risks the rules being left behind if the technology takes a new and unforeseen form.

Even if the technology can be defined with sufficient precision, regulators will need to decide what particular *types* and *uses* fall within the new rules. It's unlikely that one set of rules would be suitable for all kinds of AI; rules applicable to the riskiest forms will be overly burdensome for those that are relatively low risk. In some areas, this has led to some more targeted regulatory responses, with specific rules for specific forms and uses of AI. For instance, there are particular rules governing the use of facial recognition technology, and there may be particular rules for driverless cars. It is possible that some kinds of AI-assisted health tools to be deemed to merit their own particular body of rules.

These type- and use-based rules may even have the advantage of avoiding the need for a general-purpose definition of AI. It seems unlikely, though, that most AI healthcare tools will be regulated in such a particular way. Based on current discussions and proposals, it seems more likely that most AI healthcare tools will be regulated in a manner similar to other medical devices, with general rules for classes of the technology rather than separate rules for each individual type and use. If so, the risk associated with the use of such tools seems the most likely basis for classification.

Classifying AI healthcare tools according to risk could prove somewhat contentious. Some types and uses might seem like obvious candidates for being classified as high risk and therefore subject to stricter regulatory scrutiny and standards; AI embedded in implantable devices, or directly involved in patient treatment (robo-surgeons) might be obvious examples. What, though, of AI that has no direct patient contact, but which supports or informs human decisions rather than directly providing treatment? Choice of medications for heart diseases or chemotherapy for cancers can result in life and death situations. Should they be classified as high-risk or low risk?

Clinical Decision Support Software (CDSS) has been described by Australia's Therapeutic Goods Administration (TGA) as "software that can perform a broad range of functions that facilitate, support and enable clinical practice." [17]. Examples provided by the TGA include

- A web-based application that provides information about particular diseases or conditions based on a health practitioner's input of their patient's symptoms;
- Software intended to analyze an X-ray image to assist a radiologist in identifying anomalies; and
- Software that collects and records data from a closed-loop blood glucose monitor.

Precisely how these will fit within any given regulatory scheme will vary between jurisdictions, but a common question will be whether it qualifies as a "medical device" (or whatever analogous term is used in a particular jurisdiction). This in turn will depend on the device's intended purpose [18]. Some regulators have offered guidance on the sorts of factors that would inform such a decision. The UK's MHRA, for instance, offers the following practical examples of apps that would be considered medical devices:

- Those who calculate medicine doses for you to take/inject
- Those that tell you that you have a medical condition or disease or give you an individual percentage risk score of having one [19].

For their part, the TGA emphasizes that CDSS will not be regarded as a medical device where it is merely "digitizing what was previously a paper-based resource through an interactive software interface." Another consideration is whether the final action or decision will rest with a human practitioner. Hence, the CDSS will not be regarded as a medical device where "The health professional uses information about their patient to decide which path to proceed, and whether to use the resulting recommendation. The software itself is not directly performing any of the activities in the definition of a medical device." [17].

In other cases, the device or software is not intended to be used by medical practitioners at all, but rather, to provide information directly to layperson users. A range of wearable devices and fitness trackers is now available that can provide readings on, for example, heart rate or oxygen saturation, reminders to take medication, recommendations for exercise, and many other purposes. And as the "internet of things" continues to expand, a range of other "smart" devices could offer information and recommendations with a bearing on health.

The regulatory approach to such devices and apps thus far has been relatively light touch. US law, for instance, makes it clear that software which is intended "for maintaining or encouraging a healthy lifestyle and is unrelated to the diagnosis, cure, mitigation, prevention, or treatment of a disease or condition" is not regarded as a medical device [20]. And the FDA has made it clear that "general wellness products" perceived to present a low risk will not attract regulatory scrutiny [21].

It may be, though, that expansion of functionality of such devices and apps will require a different response. There may be no great risk if a fitness tracker is slightly inaccurate in tracking our sleep patterns or personal best 5k time. But the stakes are considerably higher if, for instance, someone with COVID-19 is using their smart-watch to monitor their oxygen saturation levels. Manufacturers' disclaimers, insisting that their products are not intended to substitute for medical advice, may not greatly reassure us if those devices are being used to determine when medical treatment is required.

Premarket regulation

Regulation of products can most obviously be divided into pre- and post-market. Many jurisdictions have regulatory checks before medical devices can be imported, sold, or used, and this will be true of AI healthcare tools as well. This may involve checks for safety and efficacy. As noted earlier, a common feature of premarket regulation is an early allocation of the product into one or other category determined by its perceived riskiness. In the United States, for example, medical devices are classified based on the level of risk into Class I (low risk), II (moderate risk), and III (high risk) [22], while a similar approach is followed in the EU [23].

A potential concern relates to the fact that the initial risk classification is often undertaken by the manufacturer or sponsor of the product. Products assessed by them as low risk will often evade much regulatory scrutiny. While this may often be uncontentious (not all AI will be particularly risky), it may amount to a vulnerability in the system when dealing with products whose risk profile is uncertain, or with risks that may not be apparent to the developer. Penalties certainly exist for knowingly misrepresenting the status of a medical device, but the complexity of the regulatory terrain that manufacturers have to navigate could lead to genuine errors of classification. While large corporate manufacturers should be able to afford expert lawyers to assist with this, the compliance costs of self-assessment for smaller developers could be problematic.

Another challenge for premarket regulation could relate to the manner in which AI healthcare device, and software as medical device (SaMD) more generally, can be acquired. Whereas traditional medical devices would require to be physically imported, AI and other forms of software can be acquired by other means—most obviously, by being downloaded, potentially from websites in other jurisdictions. The UK's MHRA is currently considering possible regulatory change, "to clarify or add to the requirements for placing SaMD on the market in these circumstances." They have proposed a change to the definition of "placing on the market," "to clarify when SaMD deployed on websites, app stores (for example, Google Play and Apple stores), and via other electronic means accessible in the UK amounts to 'placing on the market'." [3].

Postmarket regulation

Most medical devices/therapeutic products, it might be assumed, will continue to operate in much the same way after their initial approval. When they malfunction, break, or wear out, they will follow a more or less predictable pattern. This will doubtless be true of many AI tools and SaMD as well. The FDA uses the term "locked" to describe an algorithm "that provides the same result each time the same input is applied to it and does not change with use. Examples of locked algorithms are static look-up tables, decision trees, and complex classifiers." [1].

Other AI algorithms, however, are not static. In many cases, they will be capable of modification, either consciously by the user or as a product of their adaptive nature. As the FDA notes:

> not all AI/ML-based SaMD are locked; some algorithms can adapt over time. The power of these AI/ML-based SaMD lies within the ability to continuously learn, where the adaptation or change to the algorithm is realized after the SaMD is distributed for use and has 'learned' from real-world experience. [1]

Any regulatory infrastructure must be able to respond to such potential post-market changes. A "one-time check" system of premarket approval may be inadequate for these purposes. The FDA has issued a warning that regulation predicated on an assumption of a static form and use may not be well suited to all forms of AI [1]:

> The traditional paradigm of medical device regulation was not designed for adaptive AI/ML technologies, which have the potential to adapt and optimize device performance in real-time to continuously improve healthcare for patients.

This has led them to the provisional conclusion that:

> The highly iterative, autonomous, and adaptive nature of these tools requires a new, total product lifecycle (TPLC) regulatory approach that facilitates a rapid cycle of product improvement and allows these devices to continually improve while providing effective safeguards.

The FDA's consultation document considers a range of strategies, including a requirement to indicate at the initial approval stage how such adaptations will be safely managed; an obligation on manufacturers to keep such products under review; and a duty to make a new submission if the product changes beyond the intended use for which it was previously authorized.

The FDA's approach is still a work in progress and may be amended further before it comes into effect. Nonetheless, it seems likely that some form of total product lifecycle approach will be required, should such adaptive tools become part of our healthcare system. An important question, though, is how this will be carried out. Will manufacturers or owners be required to carry out regular checks and tests on their products to determine whether and how they have changed? Will regulators rely on reports from end users (in this context, presumably healthcare professionals and allied services)?

Legal obligations could also be placed on manufacturers. New Zealand's proposed new therapeutic devices law stipulates that major changes to products will see them treated as a different product, and therefore presumably subject to a fresh approval process while requiring that minor changes be notified to the regulator. The Bill also allows for use restrictions to be applied to devices. In theory, then, adequate regulatory tools might well be available to the regulator to ensure that AI algorithms are not modified or used beyond the limits of the approval granted to them. Much, though, will depend on whether the regulator is sufficiently familiar with the nature of AI and the sorts of challenges it can pose, and sufficiently resourced to be able to conduct the sorts of checks that will be needed if we are to be reassured about them.

Transparency and consent

A particularly well-documented concern about AI, in general, relates to what is often called the black box problem (Chapter 5). This inability to understand or describe exactly how an AI reaches its decisions could pose a particular issue in the context of healthcare, where it could pose challenges for informed consent. The Nuffield Council on Bioethics has raised a concern that:

> If AI systems are used to make a diagnosis or devise a treatment plan, but the healthcare professional is unable to explain how these were arrived at, this could be seen as restricting the patient's right to make free, informed decisions about their health. [24]

This has led academic commentators to raise questions about the extent to which "clinicians have a responsibility to educate the patient around the complexities of AI, including the form(s) of ML used by the system, the kind of data inputs, and the possibility of biases or other shortcomings in the data that is being used?" [25].

What level of understanding will patients need to possess about AI systems? Informed consent does not necessarily require that patients understand complex details about their conditions or treatments. It's probably rare for patients to want to know precisely why a given drug is recommended for their situation. The sorts of questions we are more likely to ask relate to possible side effects or contra-indications.

Nonetheless, there may be circumstances where some patients will want more detail. This may relate to personal risk factors:

> the practitioner will not be able to answer some very natural 'why' questions such as 'why do I have such a big risk of developing breast cancer?' that we may imagine [a patient] would have. [26]

Or it may relate to the situation where a treatment hasn't worked or has produced an unexpected side effect. In such circumstances, patients may reasonably expect an explanation for what went wrong.

Nonetheless, regulators should perhaps be wary of demanding a much higher standard from AI than from other areas of healthcare. As Tim Dare has argued: "Health professionals do not, and cannot, explain how a lot of familiar health technology works – digital thermometers; magnetic resonance imaging scanners (MRIs) ... But patients should not care." For Dare, "what matters is not transparency, or 'explainability,' but whether there is evidence of reliability." [27]. It may be that there will be instances of inevitable regulatory tension between the objectives of transparency and accuracy, but if Dare is right about the priorities of most patients, then regulators should be wary of sacrificing more accurate tools in favor of those which can explain why they have failed.

Devices or practitioners?

AI in healthcare occupies an interesting place in regulatory terms. Most current uses seem logically to fall within that part of the system that provides for medical devices, albeit the rules may need to be tweaked in certain situations. Respect to make adequate provision for some kinds of AI. But as the technology becomes more sophisticated, doubts may arise as to whether that is the appropriate regime for dealing with other kinds.

Some applications, such as conversational agents in the mental health context, will have interactions with human "healthcare consumers" that are of a very different nature from what we would typically associate with medical devices. We might in fact wonder whether these interactions are, to some extent, more appropriately managed within a regulatory paradigm designed for interactions between human healthcare providers and patients. New Zealand's Health Practitioners Competence Assurance Act 2003, for example, places a duty on the Medical Council (and other authorities specific to different healthcare roles) "to set standards of clinical competence, cultural competence (including competencies that will enable effective and respectful interaction with Māori), and ethical conduct to be observed by health practitioners of the profession." Some of this involves making sure that practitioners are suitably qualified, and that their skills and knowledge stay up to date. Others, however, relate to the manner in which doctors conduct themselves toward patients.

With regard to competence and ethical conduct, the Medical Council sets out these standards in its Good Medical Practice document. The standards in this document include requirements "to establish and maintain trust with your patients," to "make sure you treat patients as individuals and respect their dignity and privacy" and to "be courteous, respectful, and reasonable."

The Code of Health and Disability Services Consumers' Rights establishes a number of important *rights* that "healthcare consumers" are entitled to expect. Some of the Code's rights reflect concerns that we would certainly expect to be taken into account when AI is deployed in a healthcare setting; for example, "the right to have services provided with reasonable care and skill." Other rights, however, may pose questions if AIs are assuming patient-facing roles previously taken by humans. For example:

the right to be treated with respect (Right 1(1))

the right to be provided with services that take into account the needs, values, and beliefs of different cultural, religious, social, and ethnic groups, including the needs, values, and beliefs of Māori (Right 1(3))

the right to have services provided in a manner that respects the dignity and independence of the individual (Right 3)

the right to an environment that enables both consumer and provider to communicate openly, honestly, and effectively (Right 5(2))

The presence of these requirements under the 2003 Act and the Code raise a question as to whether it should be a requirement, before AI tools are deployed in consumer-facing settings, to ensure that they are able to comply with the same standards that we currently set for humans. How might such a requirement be discharged? Can a conversational agent be imbued with "cultural competence" or assured to respect the dignity of the individual?

These are questions that will surely be the subject of ongoing scrutiny in the medical ethical literature and beyond. From a regulatory perspective, though, what is interesting to note is that AI in healthcare looks likely to straddle two previously distinct strands of regulation. Insofar as it is viewed as an artifact, then it will be subject to the therapeutic products regime, oriented toward risk minimization. But in those contexts where AI performs in a more "human" way—communicating directly with healthcare consumers—then it arguably should also be evaluated against the framework that exists to ensure that human healthcare providers conduct their duties in a respectful and culturally competent manner. Whether this can be achieved within the current regulatory framework, or whether it will require some more innovative approach that fulfills these different roles, will be a matter for further study.

References

[1] U.S. Food & Drug Administration. Proposed regulatory framework for modifications to artificial intelligence/machine learning (AI/ML)-based software as a medical device (SaMD)—Discussion paper and request for feedback; 2019.

[2] U.S. Food & Drug Administration. Artificial intelligence/machine learning (AI/ML)-based software as a medical device (SaMD) action plan; 2021.

[3] Medicines & Healthcare Products Regulatory Agency, United Kingdom. Consultation on the future regulation of medical devices in the United Kingdom—Chapter 10: Software as a medical device., 2022, https://www.gov.uk/government/consultations/consultation-on-the-future-regulation-of-medical-devices-in-the-united-kingdom/chapter-10-software-as-a-medical-device. [Accessed 2 April 2023].

[4] Ministry of Health, New Zealand Government. Therapeutic products regulatory regime n.d. https://www.health.govt.nz/our-work/regulation-health-and-disability-system/therapeutic-products-regulatory-regime (Accessed 2 April 2023).

[5] Frankel S, Yeabsley J. Introduction. In: Learning from the past, adapting for the future: Regulatory reform in New Zealand. LexisNexis; 2011 [chapter 1].

[6] Moses LB. How to think about law, regulation and technology: problems with 'technology' as a regulatory target. Law Innov Technol 2013;5:1–20. https://doi.org/10.5235/17579961.5.1.1.

[7] The UK's recent Medical Devices (Amendment etc.) (EU Exit) Regulations 2020.

[8] There have already been some court judgments examining the use of AI algorithms. Loomis v Wisconsinconcerned a challenge to the use of the COMPAS algorithm in the US correctional system. It is very likely that others will follow.

[9] Human Assisted Reproductive Technology Act 2004, s 35(4). Secondary legislation refers to legislation made by someone other than the legislature, under delegated powers.

[10] Manning J. Section: Determining breach of the standard of care. In: Health law in New Zealand. Wellington: Thomson Reuters; 2015. p. 168.

[11] The World Health Organisation's guidance on Ethics and governance of artificial intelligence for health.

[12] Brownsword R, Goodwin M. Four key regulatory challenges. In: Law and the technologies of the twenty-first century: Text and materials. Cambridge, UK; New York: Cambridge University Press; 2012. p. 64.

[13] Terry NP. Of regulating healthcare AI and robots. SSRN Electron J 2019. https://doi.org/10.2139/ssrn.3321379.

[14] Levy D. Love + sex with robots: The evolution of human-robot relationships. London: Duckworth Overlook; 2009.

[15] As the House of Lords Select Committee on Artificial Intelligence noted in its 2018 report, AI in the UK: Ready, Willing and Able?: "There is no widely accepted definition of artificial intelligence. Respondents and witnesses provided dozens of different definitions".

[16] Brownsword R. The challenge of regulatory connection. In: Rights, regulation, and the technological revolution. 1st ed. Oxford: Oxford University Press; 2008. p. 160–84.

[17] Therapeutic Goods Administration, Department of Health, Australian Government. Clinical decision support software—Scope and examples; 2021.

[18] Medical Devices Regulations 2002, s.2(1) (UK); Therapeutic Goods Act 1989, s 41BD (Australia).

[19] Medicines & Healthcare Products Regulatory Agency, United Kingdom. Guidance: Medical device stand-alone software including apps (including IVDMDs); 2014.

[20] Federal food, drug, and cosmetic act, s 520, as amended by 21st century cures act (cures act).

[21] U.S. Food & Drug Administration. General wellness: Policy for low risk devices; 2016.
[22] Federal food, drug, and cosmetic act, s 513(a); Code of federal regulations, Title 21—Food and Drugs, Part 860.3.
[23] Regulation (EU) 2017/745, s 51.
[24] Nuffield Council on Bioethics. Briefing Note: Artificial intelligence (AI) in healthcare and research; 2018.
[25] Gerke S, Minssen T, Cohen G. Ethical and legal challenges of artificial intelligence-driven healthcare. In: Artificial intelligence in healthcare. Elsevier; 2020. p. 295–336.
[26] Bjerring JC, Busch J. Artificial intelligence and patient-centered decision-making. Philos Technol 2021;34:349–71. https://doi.org/10.1007/s13347-019-00391-6.
[27] Dare T. Ethics of artificial intelligence and health care. N Z Med Stud J 2019;28:5–7.

CHAPTER 9

Health inequalities in AI machine learning

Roger Yat-Nork Chung[a,b,c] **and Ben Freedman**[d,e]
[a]JC School of Public Health and Primary Care, The Chinese University of Hong Kong, Sha Tin, Hong Kong SAR
[b]CUHK Centre for Bioethics, The Chinese University of Hong Kong, Sha Tin, Hong Kong SAR
[c]CUHK Institute of Health Equity, The Chinese University of Hong Kong, Sha Tin, Hong Kong SAR
[d]Heart Research Institute, Sydney, NSW, Australia
[e]Charles Perkins Centre and Faculty of Medicine and Health, University of Sydney, Sydney, NSW, Australia

Introduction

There has recently been an increasing use of artificial intelligence (AI) in clinical care with the purposes of improving diagnosis, predicting prognosis, identifying phenotypes, selecting treatments, and increasing health system efficiency [1–3]. Despite being an object of excitement, it also instills fear for its possible ethical infringement. This chapter examines how AI machine learning may induce a particular kind of ethical issue—health inequalities. First, a review will be presented on the biases in AI machine learning processes in the typical settings of hospitals and clinics where they are designed to operate and how these biases can perpetuate and exacerbate the healthcare inequalities that are already existing. While this has been the focus of much of the literature on AI machine learning and health equity, it nevertheless represents a narrow set of challenges related to health equity. Therefore, this chapter proceeds further to bring up the less examined dimension of the potential impact of AI on the social environments in our world, which may then act as social determinants of health and healthcare inequalities.

Biases in AI machine learning

The principle in healthcare for AI machine learning is that it automatically identifies statistical patterns that are generated by hundreds of millions of data points (also known as big data) from healthcare settings to train the computer machines to perform certain specific healthcare-related tasks with precision and minimal error. A major advantage is that the computer can use all the data available that may otherwise be too cumbersome for an individual to

Artificial Intelligence in Medicine
https://doi.org/10.1016/B978-0-323-95068-8.00009-1

review in its entirety. This builds on standard statistical approaches to using registries and databases to define input variables thought important in defining risk, as a means to establish diagnoses or risk scores, but theoretically does not restrict the models to the variables that have been selected. Examples from this literature using statistical approaches provide a window into the potential pitfalls of AI machine learning for big data.

Wonderful as it may sound, machine learning has problems and biases, that may then exacerbate healthcare inequalities, much as has been shown for standard statistical models, since it relies on the quality of the data being used to train the AI models. This is the concern of the field of algorithmic fairness, and is different from, but adding on to, the potential data bias and data equity as described in Chapter 2. Here, we summarize the major potential biases across the various stages of AI machine learning, from model design to model training and prediction, deployment, and evaluation.

Biases in model design

Two primary types of bias may arise during the stage of model design, namely label bias and category bias.

In machine learning, a label is the output of the model. For instance, in a healthcare scenario, the prediction model may automatically analyze the input data, which is known as feature of the model, in order to make a prediction of the label (e.g., a diagnosis). The input features may include demographic and socioeconomic factors, images, test results and findings, clinical notes, and any other information contained in the electronic health record (EHR) of the patient, while the output label could be a detection, a diagnosis, a risk assessment, or a treatment decision. A label can be subject to bias, especially when the label itself is flawed, hence the term **label bias**. For instance, healthcare inequalities caused by systematic misdiagnoses for certain diseases for certain groups of patients will be perpetuated by AI models since they only rely on the features being inputted and must learn using the labels previously determined as correct, without the ability to distinguish whether they are biased or not. This is also known as the test-referral bias [4].

On the other hand, a **category bias** is a problem that may arise when the factor being used as a feature of the model defaults to the conventionally measured groups without taking into consideration the more granular or minority categories. Typical examples are gender being only recorded as male or female, without acknowledging other nonbinary identifications, or race being recorded as the predominant race or the nonpredominant race in the context of many nation-states. We should note that many of these variables considered in algorithmic fairness measures are socially constructed; thus, broader attention to the systemic processes involved in their construction is also necessary to put the focus on the root causes of inequity [5].

While it is important to account for social factors like race in certain analyses, using them as proxies may obscure the variations within and between these categories, and gradients within categories of these socioeconomic factors can also be overlooked. This introduces category bias. A study recently found that several clinical estimation tools across cardiology, nephrology, obstetrics, and other specialties that used simple race variables severely compromised the health of the marginalized individuals [6]. In a number of examples cited, use of Black races as part of the algorithm significantly lowered estimated risk, having the effect of discouraging physicians from undertaking further investigations and potentially delaying diagnosis and intervention.

In fact, lived experiences of individuals are always the product of intersecting patterns of different social forces, including race, occupation, education, gender, etc., which in turn calls for the need to develop new machine learning approaches that do not just rely on modeling simple additive and multiplicative effects but also consider the dynamic intersectionality of different social attributes with multilevel analytical frameworks [7].

Biases in model training and prediction

Model training problems may result from poor data. Since AI must learn to diagnose using large datasets, the model will not be reliable if the data are not representative of the different types of patients of diverse backgrounds. This is called **representative bias**. Over- or underrepresentation of certain groups could lead to biased results, because a typical machine-learning program aims to maximize overall prediction accuracy; hence, the model would optimize for the individuals that appear more frequently in the training dataset [8]. For example, it has been shown that attempts to use data from the Framingham Heart Study, a long-term, ongoing cohort study of predominantly White residents of Framingham, Massachusetts, United States, to predict the risk of cardiovascular events in non-White populations have led to biased results [9]. In particular, despite a growing number of people of mixed race/ethnicity, they were excluded from the analysis. Also, Asian and Hispanic race/ethnic groups in the study were relatively small, which contributed to a lack of predictive power. Therefore, conclusions made on these groups could be biased and should be interpreted with caution. Another example is a model of precision medicine, which is based primarily on the data of European ancestral genotypes [10]. This poses issues for machine learning of genetic associations in that 96% of participants in genome-wide association studies (GWAS) were of European descent [11].

Also, data for certain groups that are missing in a nonrandom fashion can lead to **missing data bias**. For instance, patients under certain severe conditions or isolation may have fewer vital signs, rendering any possibly critical clinical deteriorations harder to be detected. A review showed that the majority of algorithms based on electronic health records (EHRs) failed to correct for any missing data, and less than 10% corrected for all missing dimensions [12].

Moreover, some featured characteristics may make it more difficult for the AI model to generate a prediction in certain groups of patients. For example, early melanoma in darker-skinned patients may be harder to be picked up by a detection model if the informativeness of the input features is limited [13]. It has also been found that facial recognition programs had incorrectly classified more than one-third of dark-skinned women but less than 1% of light-skinned men [14].

Fig. 9.1 schematically illustrates the biases in AI model design and training, where Scenario A depicts the ideal situation where the data that

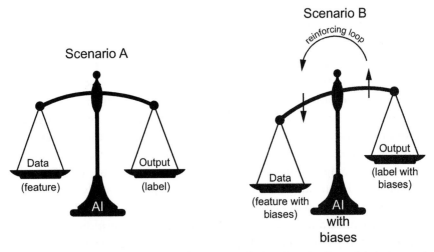

Fig. 9.1 Ideal (Scenario A) vs biased (Scenario B) situations in AI model design and training.

are inputted into the model can accurately represent the output, and Scenario B depicts the situation where the biases in the model design and training are tipping the scale, and the flawed label reinforces the already flawed AI model.

Biases in model deployment

However, even if the model design and training are free of biases, potential bias could still arise when a model is to be deployed on patients whose data characteristics and structures are not similar to those being used to train the data [1]. In other words, a unified and standardized data collection and dataset framework should be in place to avoid such issues. For biases specific to the stage of model deployment, we can categorize them into those related to clinicians or healthcare providers and those related to patients.

Biases in interaction with healthcare providers

Biases can be introduced without the awareness of the healthcare professionals when they used the algorithm. There are several of these biases. First, **automation bias** may arise when clinicians, being unaware of the potential biases that a model may have on specific groups, trust the model without adequate scrutiny of the model's integrity. This is similar to a driver trusting the automated driving system of a car all too much and failing to question it even when there is clearly danger up ahead until an accident happens. Then, if the clinician always accepts the recommendations made by the model even when the wrong output labels have been made, the model will be reinforced to perpetually repeat the same mistake due to the high matching/agreement rate between the machine recommendations and the actual human decisions.

On the other extreme, a clinician can be so aware of the systematic error that the model has been making for a certain specific group of patients that (s)he becomes desensitized or fatigued to any alert being made by the model, thereby dismissing some real cases. This would happen more often for those patients whose social groupings are not well represented in the AI model, and they are often the more socially disadvantaged individuals. All these biases may subsequently lead to more socially disadvantaged patients being inadvertently denied the healthcare resources they need.

Biases in interaction with patients

When patients intreacts with these model algorithms, biases can also be introduced inadvertently. A useful example of this bias is when AI models are not developed for those diseases that disproportionately affect the socially disadvantaged groups, these people will be affected by such bias [15]. In order to maximize positive health outcomes with limited resources in a utilitarian tradition, it is more beneficial to develop AI models for those conditions that have the greatest disease burdens in the community and not so much for conditions that are prevalent in the socially disadvantaged groups such as an ethnic minority or for conditions that are rare in prevalence. Also, there could be a mismatch of resources when models are disproportionately deployed to areas where those in need do not seek help. This will again exacerbate health inequalities if these models are deployed mainly in socially desirable areas, thereby making it inaccessible for the socially disadvantaged, consolidating the operation of the inverse care law [16], where the availability of good healthcare tends to vary inversely with the need of the population served. Of course, this presumes that the presence of AI machine learning has actually improved the quality of care.

A scoping review also found that the populations who might benefit most from innovative health technologies like optimized medical interventions (including the poor, the older persons, the rural residents, and the disabled) are the least likely to use platforms that can generate big data [17]. Related to this, there could also be avoidance behavior of patients in seeking care from clinicians and/or systems that utilize these AI models due to low level of trust in technology. Alternatively, some patients may even withhold information to influence the AI models so that they will not work against providing them with what they perceive as necessary care. Moreover, the opinions of the socially disadvantaged and the protected groups (e.g., pregnant women or children) may not be factored in during the development, utilization, and evaluation stage of the model, thereby causing **agency bias** [1]. Although these groups may be excluded from these processes for a variety of reasons, one thing that is common among them is that they lack the resources, education level, or social status to detect biases, voice out their concerns, or influence changes to be made to the model.

Biases in model evaluation

Finally, even if the socially disadvantaged patients believe that the AI models are erroneous and biased, they may not have the resources, power, and social capital to influence and bring about the necessary changes through evaluation [1]. In other words, it is important to maximize the representativeness of diverse social groups in the model evaluation stage.

Impact of AI on social determinants of health

The first part of this chapter described the direct impact of AI machine learning models on health equity. These concerns over the biases in AI model design, training, deployment, and evaluations have also been shared by the latest WHO guidance on "Ethics and Governance of Artificial Intelligence for Health" [18]. However, relatively little effort has been directed at investigating how these emerging technologies can affect health equity via their impact on the general socioeconomic, lifestyle, and cultural environments. It is imperative to consider how AI machine learning affects the social determinants of health outside the hospital and clinical settings. According to a metaanalysis in a systematic review, it is estimated that these conditions in which people are born, grow, live, work, and age [18] account for 25%–60% of mortality in any given year [19]. In fact, healthcare is only partially responsible for population and public health. Taking the social determinants of health framework of the World Health Organization (WHO) as reference (Fig. 9.2) [20], the health system is a downstream intermediary determinant of health, rather than an upstream factor. Therefore, focusing only on the impact of AI models in the healthcare settings on health

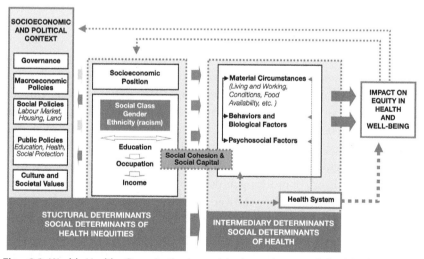

Fig. 9.2 World Health Organization's social determinants of health framework. *(Reproduced from Solar O, Irwin A., A conceptual framework for action on the social determinants of health, Social Determinants of Health Discussion Paper 2 (Policy and Practice), Executive summary, p. 6; 2010. https://www.who.int/publications/i/item/ 9789241500852, [Accessed 29/03/2022].)*

equity is only addressing part of the issue, especially since such innovative technology increasingly affects many other aspects of our lives, which interactively will affect the outcome of health in the population.

AI machine learning itself may not be considered a social determinant of health, just like technologies of internet and fiber optics are also not social determinants of health. It can, however, be a technological factor that cuts across all the social determinants of health. In other words, the question future research needs to explore is how AI machine learning is affecting the downstream, intermediary social determinants like material circumstances, behaviors, and psychosocial factors, as well as the more upstream structural social determinants of health inequities, including the socioeconomic profiling of a population and its socioeconomic and political context (e.g., governance, policies, culture, and societal values). We believe sociological and social science perspectives are needed to answer the first part of the question—i.e., how AI is shaping the socioeconomic and cultural environments before social epidemiology takes on the second part of examining the impact of this social shaping on health inequalities.

Much of the literature on AI and health inequalities has focused on the biases associated with AI model algorithms, design, training, and deployment. Only recently, some sociological efforts have been made to analyze the social shaping of AI in practice. We hereby highlight two aspects that may be particularly pertinent to the discussion on social determinants of health inequalities.

First, sociological analyses of the introduction of AI systems to workplaces have shown how their uneven uptake by users may contribute to the production of social inequalities. For instance, Brayne and Christin [21] examined an urban police department to show how their organizational contexts could be related to their workers' ability to resist the use of predictive AI analytics.

They recognized two strategies of resistance being used—(1) outright ignoring the AI and/or blocking its data collection and (2) adding more data to dilute their findings. They showed that the "enduring role of discretion, power, and occupational cultures" in the police department could shape the "social and political impact of technological change" [21]. Brayne [22,23] also challenged the common notion that more data will lead to greater justice and equity by demonstrating that the introduction of AI predictive analytics in the Los Angeles Police Department has instead deepened inequalities. She showed that AI use in fact led to more surveillance of people already considered suspicious and exacerbated spatial inequalities by subjecting low-income neighborhoods to greater assessment and intervention. These examples are relevant to the discussion on social determinants of health because people, especially those in the lower socioeconomic position communities, may be afraid of leaving digital traces that increase their visibility in surveillance systems, thereby discouraging them from seeking resources (including but not limited to health and social care) that they need. This is especially problematic when external and social data (e.g., the utilization of services, purchasing history, travel history, credit history, other neighborhood factors, etc.) are integrated into criminal surveillance systems.

Second, the impact of AI on shaping the general landscape of work is another major concern, especially as there seems to be a prevalent belief that AI will replace humans as workforce when jobs become increasingly automated. If this is what will be happening in the future, more attention needs to be paid to how a highly automated society may alter people's lifestyle behaviors, especially if their physical activities generally decrease.

Shestakofsky's participant-observation research at a software firm also found something interesting—the new AI systems were far from monolithic. In other words, the AI systems had not necessarily replaced workers but led to a new form of complementarities between technologies and workers [24]. In particular, the study revealed that the workers have taken on new roles with the introduction of AI, namely they transformed their job nature to providing "computational labor that supports or stands in for software algorithms and emotional labor aimed at helping users adapt to software systems" [24]. For that reason, we should not assume that a total labor replacement will inevitably take place but should examine the interactions between human workers and AI machines in different occupational settings and how this new form of complementarity will affect the spatial distribution of jobs and systems of inequalities. On the upside, it may take away the burden of many mundane work practices of healthcare works such as reading screening mammogram files or capsule endoscopy video. Radiologist and gastroenterologist may then spend more time to take care of the physical, social, and psychological aspects of his/her patients.

Inevitably, there will be many questions asked, mostly out of anxiety of job displacement. Will these new forms of AI-enhanced occupations sideline the conventional labor-intensive or unskilled occupations even further, perpetuating the underlying social inequalities? Are these new AI-enhanced jobs only offered in socially desirable areas? Will the education system be able to build the future workforce's capabilities to take up jobs in an AI-enhanced environment so that equal opportunity can be achieved? The real effects and impact of AI-assisted medical care are not yet known, although we can see some trace of light. What we are certain of is that all these changes will give rise to a whole new set of working environments, which in turn will act as social determinants of health, transforming inequalities of health in ways that have never happened before in human history.

Conclusion

It should be pointed out that, particularly for the social determinants of health, we have hardly scraped the surface of the multiple aspects of social conditions being shaped by AI machine learning that could serve to accentuate health inequalities, which first require recognition of potential impact and then lead to further investigations. The social determinants of health concept offer a comprehensive framework to help investigators navigate the vast diversity of potential social determinants of health that may be affected and shaped by the advent of AI machine learning technologies.

Our culture and lifestyle may never be the same with AI machine learning becoming increasingly ubiquitous in our world, shaping our sociotechnical system in a fundamental way. New AI-related laws and policies will certainly be in place for more effective governance in the near future. Therefore, it is critical for the concept of social determinants of health to be considered in making these new policies, which will have effects in shaping the material circumstances, behaviors, and psychosocial factors of people in general. While the biases in AI machine learning in healthcare settings are drawing much of the attention in gauging the potential impact on exacerbating healthcare inequalities, we should look beyond the healthcare settings to the social impact of AI if indeed we espouse to address and reduce AI-related health inequalities in the wider society (Box 9.1).

BOX 9.1 Summary of chapter.
- Despite the increase use of AI in healthcare settings, biases in AI model design, training, deployment, and evaluations can exacerbate and perpetuate health inequalities.
- Beyond the biases of AI machine learning processes, there are also impact of AI made on the social environments of our world, which may act as social determinants of health and healthcare inequalities.
- To address and reduce AI-related health inequalities in the wider society, the focus cannot only be on the healthcare settings.

Acknowledgment

The authors would like to acknowledge Mr. Alvin Hui for his assistance in the literature search, formatting of the chapter, and other technical support.

References

[1] Rajkomar A, Hardt M, Howell MD, Corrado G, Chin MH. Ensuring fairness in machine learning to advance health equity. Ann Intern Med 2018;169:866–72. https://doi.org/10.7326/M18-1990.

[2] Saria S, Rajani AK, Gould J, Koller D, Penn AA. Integration of early physiological responses predicts later illness severity in preterm infants. Sci Transl Med 2010;2. https://doi.org/10.1126/scitranslmed.3001304.

[3] Sweatt AJ, Hedlin HK, Balasubramanian V, Hsi A, Blum LK, Robinson WH, et al. Discovery of distinct immune phenotypes using machine learning in pulmonary arterial hypertension. Circ Res 2019;124:904–19. https://doi.org/10.1161/CIRCRESAHA. 118.313911.

[4] Owens DK, Sox HC. Biomedical decision making: probabilistic clinical reasoning. In: Biomedical informatics: Computer applications in health care and biomedicine. 4th ed. London: Springer; 2014. p. 67–107. https://doi.org/10.1007/978-1-4471-4474-8_3.

[5] Yearby R. Structural racism and health disparities: reconfiguring the social determinants of health framework to include the root cause. J Law Med Ethics 2020;48:518–26. https://doi.org/10.1177/1073110520958876.

[6] Vyas DA, Eisenstein LG, Jones DS. Hidden in plain sight-reconsidering the use of race correction in clinical algorithms. N Engl J Med 2020;383:874–82. https://doi.org/ 10.1056/NEJMms2004740.

[7] Evans CR, Williams DR, Onnela J-P, Subramanian SV. A multilevel approach to modeling health inequalities at the intersection of multiple social identities. Soc Sci Med 2018;203:64–73. https://doi.org/10.1016/j.socscimed.2017.11.011.

[8] Zou J, Schiebinger L. AI can be sexist and racist—it's time to make it fair. Nature 2018;559:324–6. https://doi.org/10.1038/d41586-018-05707-8.

[9] Gijsberts CM, Groenewegen KA, Hoefer IE, Eijkemans MJC, Asselbergs FW, Anderson TJ, et al. Race/ethnic differences in the associations of the Framingham risk factors with carotid IMT and cardiovascular events. PLoS ONE 2015;10. https://doi.org/ 10.1371/journal.pone.0132321.

[10] Popejoy AB, Ritter DI, Crooks K, Currey E, Fullerton SM, Hindorff LA, et al. The clinical imperative for inclusivity: race, ethnicity, and ancestry (REA) in genomics. Hum Mutat 2018;39:1713–20. https://doi.org/10.1002/humu.23644.

[11] Need AC, Goldstein DB. Next generation disparities in human genomics: concerns and remedies. Trends Genet 2009;25:489–94. https://doi.org/10.1016/j.tig.2009.09.012.

[12] Goldstein BA, Navar AM, Pencina MJ, Ioannidis JPA. Opportunities and challenges in developing risk prediction models with electronic health records data: a systematic review. J Am Med Inform Assoc 2017;24:198–208. https://doi.org/10.1093/jamia/ ocw042.

[13] Adamson AS, Smith A. Machine learning and health care disparities in dermatology. JAMA Dermatol 2018;154:1247–8. https://doi.org/10.1001/jamadermatol.2018.2348.

[14] Buolamwini J, Gebru T. Gender shades: intersectional accuracy disparities in commercial gender classification. In: Proceedings of the 1st Conference on Fairness; 2018. p. 81.

[15] Veinot TC, Mitchell H, Ancker JS. Good intentions are not enough: how informatics interventions can worsen inequality. J Am Med Inform Assoc 2018;25:1080–8. https:// doi.org/10.1093/jamia/ocy052.

[16] Tudor HJ. The inverse care law. Lancet 1971;297:405–12. https://doi.org/10.1016/S0140-6736(71)92410-X.

[17] Weiss D, Rydland HT, Øversveen E, Jensen MR, Solhaug S, Krokstad S, et al. Innovative technologies and social inequalities in health: a scoping review of the literature. PLoS ONE 2018;13, e0195447. https://doi.org/10.1371/journal.pone.0195447.

[18] WHO regional office for south-East Asia. Social Determinants of Health; 2008.

[19] Heiman HJ, Artiga S. Beyond health care: the role of social determinants in promoting health and health equity. Health 2015;20:1–10.

[20] Solar O, Irwin A. A conceptual framework for action on the social determinants of health. In: Social determinants of health discussion paper 2 (policy and practice); 2010.

[21] Brayne S, Christin A. Technologies of crime prediction: the reception of algorithms in policing and criminal courts. Soc Probl 2021;68:608–24. https://doi.org/10.1093/socpro/spaa004.

[22] Brayne S. Big data surveillance: the case of policing. Am Sociol Rev 2017;82:977–1008. https://doi.org/10.1177/0003122417725865.

[23] Brayne S. Predict and surveil: Data, discretion, and the future of policing. United States: Oxford University Press; 2020. https://doi.org/10.1093/oso/9780190684099.001.0001.

[24] Shestakofsky B. Working algorithms: software automation and the future of work. Work Occup 2017;44:376–423. https://doi.org/10.1177/0730888417726119.

CHAPTER 10

Human-machine interaction: AI-assisted medicine, instead of AI-driven medicine

René F. Kizilcec[a], Dennis L. Shung[b], and Joseph J.Y. Sung[c]

[a]Department of Information Science, Cornell University, Ithaca, NY, United States
[b]Yale School of Medicine, New Haven, CT, United States
[c]Lee Kong Chian School of Medicine, Nanyang Technological University, Singapore, Singapore

Introduction

Clinical Decision Support Systems (CDSSs) have been used in medicine since the 1980s to improve healthcare delivery by assisting clinicians in complex decision-making processes that have historically relied on human cognition [1]. They take clinical knowledge, patient information, and other health information as inputs and provide a patient-specific assessment such as a risk score. Clinicians combine their own domain knowledge and experience with the assessment from a CDSS at the point-of-care to make a decision in a diagnostic workup, possible treatment plan, or level of healthcare utilization. CDSS can help clinicians, patients, and other medical stakeholders to enhance clinical decision-making with the hope of ultimately improving patient clinical outcomes.

CDSSs are computerized systems that rely on clinical data, such as the electronic health record or similar medical data infrastructures. While early CDSSs were designed as knowledge-based systems that represented deterministic rules, today's CDSSs use algorithms to model decisions without an explicitly defined knowledge base. The field of medical informatics has developed increasingly sophisticated and accurate data-driven models that can make predictions of patient outcomes using artificial intelligence (AI). Simultaneously, these systems have also become increasingly opaque or "black box." While the recommendation of a knowledge-based CDSS could be audited to reveal a specific set of rules that generated it, the recommendations of AI–CDSS, especially modern ones that use complex machine learning-based algorithms (e.g., deep neural network), cannot be explained in the same way. Some have argued it may not be necessary to open black

Artificial Intelligence in Medicine
https://doi.org/10.1016/B978-0-323-95068-8.00010-8

box models so long as they are highly accurate [2]. However, AI-CDSSs are primarily used in medicine to assist rather than drive clinical care decisions, so even when they are highly accurate, their adoption and effective use by medical professionals relies on the interaction between human and machine. In high-stakes and time-constrained medical settings, human experts may be hesitant to trust a black box model, especially if it cannot explain a recommendation that deviates from the inclination of the human expert (i.e., expectation violation).

Major advances in AI methods have accelerated efforts in clinical informatics to develop increasingly sophisticated AI-CDSSs that feature state-of-the-art deep learning. A comprehensive review of AI-CDSS conducted in 2021 found 13 FDA-approved devices under the software as medical devices (SaMD) category [3]. Many of these systems are demonstrably about as accurate, if not more, at detecting diseases or predicting the likelihood of near-term complications as human medical experts. From a bioethics standpoint, if a system offers improvements in patient outcomes, the principle of beneficence demands that it be used in clinical practice. A key area where AI-CDSS can enhance human decision-making is the prioritization of patients (i.e., triage decisions). Accurate triage decisions produce an optimal allocation of scarce resources, which can, in many cases, improve social equity in clinical care decisions. A classic example of AI-CDSS use for triage is in the context of neuroradiology, where the system helps detect bleeding in the brain of patients who present with signs of confusion [4]. FDA-approved in 2018, the Aidoc system analyzes noncontrast CT scans of the brain to detect cases of intracranial hemorrhage (ICH). The use of this particular system, among several others, is even financially incentivized by Medicare [5]. This highlights the importance of understanding how AI-CDSSs are being used effectively and where their pitfalls and potential harms lie [1].

In November 2022, the launch of ChatGPT has brought another groundbreaking advancement for AI in Medicine. There are already ad hoc cases showing that ChatGPT is able to make clinical diagnosis of individual patients based on clinical features and test results. ChatGPT can also be trained to prepare clinical notes from doctors who dictate their history-taking and physical examination of their patients. The report can clearly identify the chief symptom, ongoing complaints, and progress of disease in these patients. Would ChatGPT replace or displace physicians consultations and drive the treatment of medical conditions, at least in simple cases, without the input of healthcare providers?

Implementation is the challenge

While the development of AI-CDSSs is advancing rapidly, there are important concerns about their effective implementation in clinical practice. Prediction models are evaluated based on how well they perform on the specific task they were trained on, for example, classifying a patient as high- or low-risk to inform a triage decision. But even a model that is highly accurate, with low false-positive and false-negative rates, can fail to provide improvements in patient outcomes if it is not used effectively once embedded in a clinical setting. To evaluate the impact of an AI-CDSS on patient outcomes, we need to examine it in the context where it is used by clinicians to make decisions. The most rigorous evaluation design is a randomized clinical trial (RCT), which can provide causal evidence of the impact of integrating a system into a medical context. As of 2022, there are only 41 published RCTs of how embedding AI in medicine affects patient outcomes, but 30 out of the 41 RCTs (77%) showed that AI-assisted interventions outperformed usual clinical care [6]. Nevertheless, clinicians, patients, and policymakers have raised serious concerns about data sharing, standards, privacy, and model transparency for AI systems in medicine [7], which can create resistance that impedes their effective use.

One of the most significant concerns about AI in medicine and other high-stakes domains is the potential for bias in models that are learned from training data that reflect current and historical inequities in the provision of clinical care [8]. AI models could inadvertently perpetuate and even exacerbate forms of bias and discrimination that are encoded in the training data [9]. In order to increase awareness of potential biases from the absence of representation for underrepresented minorities, the trials studying these algorithms should report the proportion of enrolled participants that identify with underrepresented minority groups. This has been reported in only 27% of all RCTs of AI interventions in medicine, with a median of 21% participants from underrepresented groups [6]. The resulting disparate impact of the AI system risks violating the bioethics principles of nonmaleficence and justice. Considering how frequently AI systems are featured in the news for their apparent flaws and biases, it is only reasonable for clinicians to be cautious and, at times, reluctant to use them unless they are necessary. AI-CDSS will need to be perceived as unbiased, competent, easy to use, and trustworthy to encourage widespread adoption. This ideal is reflected in a recent synthesis of a range of initiatives to establish ethical principles for the adoption of socially beneficial AI [10]. Beyond the four core bioethics principles of beneficence, nonmaleficence, autonomy, and justice, the synthesis identified explicability as a fifth core ethics principle for AI systems. The explicability of an AI system demands both its intelligibility (i.e., accessible information about how it works) and accountability (i.e., information on who is responsible for how it works).

Human-machine interaction

We have identified the effective integration of AI-CDSS into clinical practice as one of the critical hurdles for realizing the potential benefits of AI in medicine. Overcoming this hurdle will require focused attention on the human-machine interaction from a behavioral and psychological perspective. The dominant interaction paradigm today is AI-assisted instead of AI-driven medicine. The AI system is there to empower a human stakeholder in making medical decisions, not take over medical decision-making for them. To design systems for successful human-machine interaction, we need a robust understanding of the people who will be using the systems, their prior knowledge and beliefs, and their goals and values to gain insights into why they would or would not use them.

We took a closer look at which stakeholders are empowered by current work on AI in medicine to understand who is primarily interacting with AI systems. We reviewed 41 RCTs of AI in medicine published between 2015 and 2021 [6] and found that all of the AI systems tested followed a human-in-the-loop design (i.e., AI-assisted medicine where AI models inform human decision-making). In the vast majority of RCTs, the AI system was targeted at direct care providers (70.7%), but in some cases, the AI system was designed for use by patients (22.0%) or indirect care providers (9.8%) [$X^2(2) = 38$, $P < .001$]. One RCT evaluated a system designed for use by both direct care providers and patients; we counted this study once for each stakeholder. We also found that the effectiveness of the AI system, as evaluated in the RCT, varied significantly based on the target stakeholder [$X^2(2) = 7.1$, $P = .029$]. AI systems for direct care providers were most likely to show a positive impact (79%), followed by AI systems for indirect care providers (50%) and patients (33%). Our review identified direct care providers as the most common target user group of AI systems and the group for which these systems are most effective. We will therefore focus on direct care providers in the remainder of this chapter. Future developments of AI in medicine may focus more on patients or indirect care providers, but it will be important to first better understand why the small number of existing systems designed for these user groups has been largely ineffective at improving patient outcomes.

The successful integration of AI-CDSS into existing practices requires a socio-technical understanding of clinician providers and their mental model of the system, which includes their understanding of how it works, the reasons it was created, and ways it can support them. The benefits of AI systems are contingent on successfully integrating them in existing practice, which is largely a nontechnical challenge; specifically, it is largely independent of the AI model's performance or even algorithmic bias. Instead, it is a social-psychological, cultural, and organizational challenge that hinges on the clinician's beliefs, perceptions, attitudes, and behaviors toward the AI system. A series of interviews with clinicians about their attitudes toward AI systems found that clinicians prefer these systems to be partners in the decision-making process, instead of surrogate decision-makers [11]. The interviews also found that clinicians can build trust in AI systems even without a full understanding of how they work. Two important mechanisms for building trust that were identified are: (1) gaining experience using the system over time and (2) seeing the endorsement and validation of the system by experts in the field. Additionally, clinicians expressed a strong preference for system designs that accommodate their autonomy and support them across their clinical workflow.

AI systems raise a novel challenge when designing for successful human-machine interaction because algorithm aversion (i.e., a negative attitude toward using algorithms) influences how people respond to AI systems in real-world environments. Studies on algorithm aversion have found that people generally trust humans more than algorithms and follow human recommendations more, especially for tasks that are subjective or require attention to individual uniqueness [12,13]. Most clinical care decisions indeed require close attention to individual patient characteristics, emotions, and socioeconomic conditions, which increases the likelihood of algorithm aversion. This can lead to irrational responses to algorithms, such as punishing them more for a mistake compared to a human who commits the same mistake. In other words, people are less likely to give algorithms the benefit of the doubt when they err; instead, they tend to view them as less competent. Along the same lines, research on algorithm appreciation (i.e., a positive attitude toward using algorithms) has shown that novices are more likely than experts to adhere to algorithmic recommendations [14], and increasing the prediction task's perceived objectivity or the algorithm's perceived human-likeness can increase algorithm use [15]. This highlights the challenge of improving attitudes toward AI-CDSSs, which are used by clinicians, who are experts, for complex clinical care decisions, which clinicians may not perceive as purely objective. AI systems in medicine are therefore at risk of algorithmic aversion without enjoying the benefits of algorithmic appreciation, unless their framing and design strategically present them to clinicians as team members. The interaction of human and machine as a team in the decision-making process would make AI much more acceptable to both patients and healthcare providers.

AI-CDSSs as clinical team members

The clinical care process has become increasingly collaborative: teams of clinicians with varying levels of experience and areas of expertise work together to provide the best possible patient care. This presents a structural design opportunity for integrating AI-CDSS into the clinical team to realize the potential of AI-assisted medicine. The opportunity for adopting a human-machine teaming paradigm came out in a series of interviews with clinicians [11]. As a "team member," the AI-CDSS would, for instance, refrain from presenting its recommendation as early as possible, before other team members have had an opportunity to review the patient information. Instead, it would offer its input during the decision-making process to support the team's effort to arrive at an optimal clinical care decision. Moreover, as a "team member," the AI-CDSS would be able to provide an explanation for its recommendation upon request. The explainability of AI models, which represents an important step toward satisfying the bioethics principle of explicability [10], is expected to promote the adoption and effective use of AI-CDSS.

An AI-CDSS can be considered explainable if it satisfies the following criteria: (1) it can justify its recommendations to clinicians in terms of what information it uses, ignores, and what it "knows"; (2) it can justify inaccuracies to clinicians with sufficient information to help them understand why it underperformed and allow them to rationalize it; and (3) it sets clear expectations for clinicians about when it works well and where its limitations lie [16].

AI-CDSS systems that incorporate explainability features in their design can elicit more positive emotional responses and perceptions of trustworthiness from the clinicians who use the systems and the patients who are affected by them [17,18]. While promoting trust in AI-CDSS is desirable, the goal should not be to maximize clinician trust in the system, because this can result in an overreliance on the AI system and an abrupt loss of trust when a problem arises. In high-stakes scenarios, explanations are an effective way to reduce overreliance on AI systems in the decision-making process because they can reveal flaws in logic or other missteps that led to an inaccurate recommendation [19]. Clinician trust in AI-CDSS will need to be developed gradually as they gain experience using it, just as a new team member has to earn the trust of their peers over time. Offering clear explanations to justify recommendations can facilitate the process of developing trust, but common explainable AI approaches that present a full explanation at every decision point may not be suitable for clinical settings [20]. Working in high-stakes, time-constrained settings, many clinicians prefer to determine upfront if a new AI-CDSS is trustworthy, and instead of receiving full explanations of complex models, they prefer a certain level of interpretability or comprehensibility of system recommendations [21].

The influence that the design of AI-CDSS has on their adoption and effective use has led to a "rise in design science in clinical research" [22]. This has included studies on how to design for AI interpretability, fairness, effective visualization, accuracy of risk communication, and more [23]. A recent co-design study elicited clinicians' perceptions of how to design an AI-CDSS for antidepressant treatment decisions [20] and identified three major design lessons. First, when the AI model output contrasts with the existing domain knowledge of the clinician (i.e., expectation violation or conflicting information), it causes confusion and calls for an on-demand explanation for what happened (i.e., explainability, transparency, and interpretability).

Second, designers should not assume a single-user system but consider how the system will be used collaboratively, for instance, by attending physicians, residents, advanced nurse practitioners, nurses, and patients.

Finally, the system design needs to consider the nature of the context, including how long and frequent the clinical interactions are and the willingness or ability of patients to participate in the decision process. Similar co-design studies for other types of AI-CDSS may provide valuable design guides. A notable limitation of most AI-CDSS design studies to date is that they rely on clinical scenarios with low levels of realism, for example, presenting clinicians with screenshots or simple prototypes of AI-CDSS in an online survey or a quiet laboratory setting without any stakes. It raises questions about how well insights from these design studies will translate to the reality of noisy and fast-paced clinical decision-making.

Positioning AI-CDSS as a member of the clinical team provides several benefits but also raises a new set of questions [24]. It is relatively simple to frame the AI system as a team member in ways that influence clinicians' perceptions and attitudes toward the technology because people treat computers as social actors [25] and respond to computer teammates in similar ways as human teammates [26]. This offers designers of AI-CDSS a straightforward heuristic to follow: design the system just like an expert human member of the clinical team would act, communicate, and respond to questions from other team members. This heuristic can help answer fundamental questions in the design of AI-CDSS, including when should AI-CDSS present recommendations and to whom, how should recommendations be communicated, what information should be provided to support the recommendation, and what additional justification and information should be provided if asked. Nevertheless, the team-member framing of AI-CDSS also raises a new set of questions related to its position in the medical hierarchy, who gets to consult with it and at what point in time, and even how it can contribute to effective interpersonal cooperation, communication, and cohesion among human team members [27].

Conclusion

In this chapter, we examined the implications of a growing number of AI systems being deployed in clinical settings to improve patient care. AI-CDSSs use opaque models learned from large amounts of data to give recommendations to clinicians, which they can consider when making clinical care decisions. The technological development of clinical AI models is outpacing the design science to guide the effective integration of AI-CDSS into clinical workflows. While there are only a limited number of AI-CDSS integrations to date, most of them have targeted direct care providers, the group for which these systems are also most effective according to the same set of RCTs. Yet, many AI-CDSSs fail to gain traction or are not used effectively because clinicians perceive them as not useful, cumbersome, or untrustworthy. Taking a human-machine interaction perspective, future development which look at social-psychological, cultural, and organizational factors that influence people's perceptions, attitudes, and behaviors toward AI and their implications in medicine. Specifically, clinicians and AI engineers should explore the potential of explainable AI techniques to support the gradual development and prevent the loss of clinician trust in AI-CDSS. To address the sociotechnical challenge of designing AI systems that are embedded in current clinical workflows instead of disrupting them, we explore the potential of framing the AI system as another member of the clinical team. Based on this framing, we develop a useful heuristic (i.e., ask what would an expert team member do?) to guide the design and integration of AI in medicine.

As an active area of research and development, we expect to see more collaboration between the fields of medical information and human-algorithmic interaction in the coming years to make significant advances in the design and implementation of AI-CDSS. By optimizing the human-algorithmic interaction and integrating the AI-CDSS as a team member with valuable input but inherent limitations, we look forward to a future where human-AI teams work collaboratively to improve patient outcomes.

We believe that AI alone should not drive clinical decision and implement therapy for patients. AI-assisted Medicine, which is more acceptable to patients and healthcare providers, is a better way forward.

References

[1] Sutton RT, Pincock D, Baumgart DC, Sadowski DC, Fedorak RN, Kroeker KI. An overview of clinical decision support systems: benefits, risks, and strategies for success. NPJ Digit Med 2020;3:17. https://doi.org/10.1038/s41746-020-0221-y.

[2] Holm EA. In defense of the black box. Science 2019;364:26–7. https://doi.org/10.1126/science.aax0162.

[3] Zhu S, Gilbert M, Chetty I, Siddiqui F. The 2021 landscape of FDA-approved artificial intelligence/machine learning-enabled medical devices: an analysis of the characteristics and intended use. Int J Med Inform 2022;165:104828. https://doi.org/10.1016/j.ijmedinf.2022.104828.

[4] Kaka H, Zhang E, Khan N. Artificial intelligence and deep learning in neuroradiology: exploring the new frontier. Can Assoc Radiol J 2021;72:35–44. https://doi.org/10.1177/0846537120954293.

[5] Seyam M, Weikert T, Sauter A, Brehm A, Psychogios M-N, Blackham KA. Utilization of artificial intelligence–based intracranial hemorrhage detection on emergent noncontrast CT images in clinical workflow. Radiol Artif Intell 2022;4:e210168. https://doi.org/10.1148/ryai.210168.

[6] Plana D, Shung DL, Grimshaw AA, Saraf A, Sung JJY, Kann BH. Randomized clinical trials of machine learning interventions in health care: a systematic review. JAMA Netw Open 2022;5:e2233946. https://doi.org/10.1001/jamanetworkopen.2022.33946.

[7] He J, Baxter SL, Xu J, Xu J, Zhou X, Zhang K. The practical implementation of artificial intelligence technologies in medicine. Nat Med 2019;25:30–6. https://doi.org/10.1038/s41591-018-0307-0.

[8] Cho MK. Rising to the challenge of bias in health care AI. Nat Med 2021;27:2079–81. https://doi.org/10.1038/s41591-021-01577-2.

[9] Barocas S, Selbst AD. Big data's disparate impact. SSRN Electron J 2016. https://doi.org/10.2139/ssrn.2477899.

[10] Floridi L, Cowls J, Carta S. A unified framework of five principles forin society. In: Machine learning and the City. 1st ed. Wiley; 2022. p. 535–45.

[11] Henry KE, Kornfield R, Sridharan A, Linton RC, Groh C, Wang T, et al. Human–machine teaming is key to AI adoption: clinicians' experiences with a deployed machine learning system. NPJ Digit Med 2022;5:97. https://doi.org/10.1038/s41746-022-00597-7.

[12] Dietvorst BJ, Simmons JP, Massey C. Algorithm aversion: people erroneously avoid algorithms after seeing them err. J Exp Psychol Gen 2015;144:114–26. https://doi.org/10.1037/xge0000033.

[13] Longoni C, Bonezzi A, Morewedge CK. Resistance to medical artificial intelligence. J Consum Res 2019;46:629–50. https://doi.org/10.1093/jcr/ucz013.

[14] Logg JM, Minson JA, Moore DA. Algorithm appreciation: people prefer algorithmic to human judgment. Organ Behav Hum Decis Process 2019;151:90–103. https://doi.org/10.1016/j.obhdp.2018.12.005.

[15] Castelo N, Bos MW, Lehmann DR. Task-dependent algorithm aversion. J Mark Res 2019;56:809–25. https://doi.org/10.1177/0022243719851788.

[16] Tonekaboni S, Joshi S, McCradden MD, Goldenberg A. What clinicians want: Contextualizing explainable machine learning for clinical end use; 2019. https://doi.org/10.48550/ARXIV.1905.05134.

[17] Dodge J, Liao QV, Zhang Y, Bellamy RKE, Dugan C. Explaining models: an empirical study of how explanations impact fairness judgment. In: IUI '19: 24th international conference on intelligent user interfaces; 2019. p. 275–85. https://doi.org/10.1145/3301275.3302310.

[18] Kizilcec RF. How much information?: effects of transparency on trust in an algorithmic interface. In: CHI'16: CHI conference on human factors in computing systems; 2016. p. 2390–5. https://doi.org/10.1145/2858036.2858402.

[19] Vasconcelos H, Jörke M, Grunde-McLaughlin M, Gerstenberg T, Bernstein MS, Krishna R. Explanations can reduce overreliance on AI systems during decision-making. Proc ACM Hum Comput Interact 2023;7:1–38. https://doi.org/10.1145/3579605.

[20] Jacobs M, He J, Pradier MF, Lam B, Ahn AC, TH MC, et al. Designing AI for trust and collaboration in time-constrained medical decisions: a sociotechnical lens. In: CHI '21: CHI conference on human factors in computing systems; 2021. p. 1–14. https://doi.org/10.1145/3411764.3445385.

[21] Rudin C, Chen C, Chen Z, Huang H, Semenova L, Zhong C. Interpretable machine learning: fundamental principles and 10 grand challenges. Stat Surv 2022;16. https://doi.org/10.1214/21-SS133.

[22] Arnott D, Pervan G. A critical analysis of decision support systems research revisited: the rise of design science. J Inf Technol 2014;29:269–93. https://doi.org/10.1057/jit.2014.16.

[23] Yang Q, Steinfeld A, Zimmerman J. Unremarkable AI: fitting intelligent decision support into critical, clinical decision-making processes. In: CHI '19: CHI conference on human factors in computing systems; 2019. p. 1–11. https://doi.org/10.1145/3290605.3300468.

[24] Sebo S, Stoll B, Scassellati B, Jung MF. Robots in groups and teams: a literature review. Proc ACM Hum Comput Interact 2020;4:1–36. https://doi.org/10.1145/3415247.

[25] Reeves B., Nass C.I. The media equation: how people treat computers, television, and new media like real people and places. 1 paperback ed., [reprint.]. Stanford, CA: CSLI Publ;.

[26] Nass C, Fogg BJ, Moon Y. Can computers be teammates? Int J Hum Comput Stud 1996;45:669–78. https://doi.org/10.1006/ijhc.1996.0073.

[27] Jung MF, Martelaro N, Hinds PJ. Using robots to moderate team conflict: the case of repairing violations. In: HRI '15: ACM/IEEE international conference on human-robot interaction; 2015. p. 229–36. https://doi.org/10.1145/2696454.2696460.

Index

Note: Page numbers followed by *f* indicate figures, *t* indicate tables, and *b* indicate boxes.

A
Accountability, 22, 34, 41, 53–54
AI-CDSS. *See* Clinical decision support system (CDSS)
Aidoc system, 132
Algorithm
  aversion, 135–136
  definition, 13
  error, 14, 17
  machine learning (ML)
    challenges, 64
    fairness in, 67
    interpretability, 65–66
  mitigation strategy (*see* Algorithmic stewardship)
  output, 22
  neural network algorithm, 1
  performance evaluation, 18
  provider bias, 20
  vigilance, 23–24
Algorithmic stewardship, 23–24
Algorithm vigilance, 23–24
All of Us Research Program, 17–18, 17*f*
Ambient sensors, 21
Application ("App")
  medical futility, 38–39
  procedural ethical issues, identification of
    accountability, 41
    engagement, 41
    reflexivity, 40
    transparency, 41
    trustworthiness, 41
  respect for persons, 41–42
  substantive ethical values, identification of
    beneficence, 39
    harm minimization, 39
    integrity, 39–40
    justice, 40
    liberty/autonomy, 40
    privacy/confidentiality, 40
    public benefit, 40
Apportioning liability, 93–94
Artificial intelligence (AI)

Application ("App")
  problems, identification of, 38–39
  procedural ethical issues, identification of, 40–41
  respect for persons, 41–42
  substantive ethical values, identification of, 39–40
  applied, 83
  challenges, 8–9
  ChatGPT, 6–7
  clinical decision support system (CDSS), 28–29
  in clinical medicine
    barriers of, 81–83
    cardiology, 77, 78*b*
    chronic disease management, 73–74, 74*b*
    continuous quality improvement (CQI), 71–72
    future research, 83
    mental health and psychiatry, 75–77, 77*b*
    need for, 79–80, 81*f*
    radiology, 74–75, 76*b*
    robotics and rehabilitation, 78–79, 79*b*
  clinicians and healthcare provider, 9–11
  creative destruction, 2
  definition, 72–73
  drug and diagnostic test discovery, 5–6, 6*f*
  ethical decision-making framework
    accountability, 34
    articulating issues, 29
    beneficence, 32
    engagement, 34
    harm minimization, 32
    integrity, 32
    justice, 33
    liberty/autonomy, 32
    potential actions, identification of, 30
    privacy/confidentiality, 33
    public benefit, 33
    reflexivity, 33
    relevant ethical merit, 30

Artificial intelligence (AI) *(Continued)*
  relevant values, identification of,
      29–30, 30*f*
  robust justifications, 31
  transparency, 34
  transparent communication, 31, 31*f*
  trustworthiness, 34
  ethical principles, 27–28
  guidance documents, 27
  in healthcare, 71
  health conditions and treatment
      outcome, 4
  image-based analysis and diagnosis, 2–3
  implementation of, 8, 10–11, 11*f*, 83
  machine learning (ML), 1
    advantage, 119–120
    bias (*see* Bias and machine learning
        (ML))
    big data, 119–120
    principle, 119–120
  in medicine (*see* Medical artificial
      intelligence (AI))
  neural network algorithms, 1
  patient's perspectives, 8
  precision medicine, 3–4
  provider decision support systems (PDSS)
    accountability, 41
    autonomy, 40
    beneficence, 39
    engagement, 41
    harm minimization, 39
    integrity, 39–40
    justice, 40
    liberty, 40
    medical futility, 38–39
    public benefit, 40
    reflexivity, 40
    respect for persons, 41–42
    transparency, 41
    trustworthiness, 41
  regulation in healthcare
    Advisory Committee on Assisted
        Reproductive Technology
        (ACART), 106–107
    applications, 114
    black box problem, 113
    challenges, 108
    clinical decision support system
        (CDSS), 109–110
    consumers' rights, 115
    court judgments, 106
    definition, 106
    guidelines, 107
    healthcare tools, 109
    informed consent, 113
    legal rules, 106
    legislation, 106
    level of risk, 111
    medical device, 109–110
    postmarket, 111–113
    premarket, 111
    smart devices, 110
    software as medical device (SaMD),
        111–112
    Therapeutic Goods Administration
        (TGA), 109–110
    transparency and consent, 113–114
  rehabilitation tools, 5
  respect, 28
    attitudinal stance, 36
    autonomy, 36
    classifications, 35–36
    definition, 34–35
    obligations, 36–37
  Science, Health and Policy-relevant
      Ethics in Singapore (SHAPES)
      Initiative, 28
  science of, 83
  ultrasound service, 5
Automation bias, 123

B
Bias and machine learning (ML)
  category, 120–121
  impact of
    health equity, 125–126
    job displacement, 127–128
    social shaping, 126
    surveillance systems, 126–127
    total labor replacement, 127
    World Health Organization (WHO),
        125–126, 125*f*
  interaction with
    healthcare providers, 123

patients, 123–124
label, 120
missing data, 122
model design and training, 122–123, 122f
in model evaluation, 124
model training problems, 121–122
prediction, 122
race/ethnicity, 121–122
representative, 121–122
social determinants, 125–128, 125f
test-referral, 120
Black box medicine
data collection, 60
disadvantages, 59
ethical issues, 61f, 68
ethical principles, 60–61f, 68
Federated Learning, 63
models and algorithm development
beneficence, 66–67
fairness, 67
interpretability, 65–66
justice, 67
nonmaleficence, 66–67
transparency, 64–65
polyp detection systems, 62–63
principles, 59
privacy of data, 62–63
transparency, 61–62

algorithm aversion, 135–136
challenges, 135
randomized clinical trial (RCT),
134–135
trust building mechanism, 135
patient-specific assessment, 131
randomized clinical trial (RCT), 133
regulation in healthcare, 109–110
team members, 136
Clinical medicine
aging population, 72
barriers of
adoption and scaling, 81–82
clinical evidence, 82–83
biomarkers, 79–80
cardiology, 77, 78b
chronic disease management, 73–74, 74b
clinical outcomes, 80, 81f
continuous quality improvement (CQI),
71–72
future research, 83
graphics processing unit (GPU), 80
mental health and psychiatry, 75–77, 77b
radiology, 74–75, 76b
robotics and rehabilitation, 78–79, 79b
Confidentiality, 47–49
Continuous quality improvement (CQI),
71–72

C
Category bias, 120–121
CDSS. *See* Clinical decision support system
(CDSS)
ChatGPT, 6–7, 132
Clinical decision support system (CDSS),
28–29
Aidoc system, 132
artificial intelligence (AI) systems,
131–132
bias, 133–134
ChatGPT, 132
clinical data, 131–132
clinician trust, 136–137
design of, 137–138
ethical principles, 133–134
human-machine interaction

D
Data access, 15f
definition, 14
error, cause of, 14
governance of data, 15–16
need for, 13–14
for research use, 17–18, 17f
security and privacy, 16–17
Data access request (DAR), 53
Data bias, 15f
algorithmic stewardship, 23–24
definition, 14
electronic health record data, 19
missing data, 19
need for, 13–14
physician-patient relationship, 18
practice patterns, 20

Data bias *(Continued)*
  provider bias, 20
  racial correction, 20–21
  wearables and ambient sensors, 21
Data equity, 15*f*
  algorithmic stewardship, 23–24
  definition, 14
  explainability, 22
  impartiality, 22–23
  inclusion, 23
  interpretability, 22
  need for, 13–14
  transparency, 22
Data protection principles (DPP)
  access and rectification, 53
  accuracy, 51
  data security breaches, 53
  erasure, 51
  explainability, 50–51
  fairness, 50–51
  lawfulness, 50–51
  organizational measures, 53
  purpose limitation and data
    minimization, 51
  retention limitation, 51–52
  right to rectification, 51
  security safeguards, 52
  technical measures, 53
  transparency, 50–51
  use and disclosure, 52
Digital data ethics
  auditability, 57
  ethical considerations, 57
  med-health sciences and services, 56–57
  profiling, 57
DPP. *See* Data protection principles (DPP)

E
Electronic health record (EHR)
  data bias, 19
  missing data, 19
  predictions, 77, 77*b*
  provider bias, 20
  output label, 120
Ethical decision-making framework
  accountability, 34
  articulating issues, 29

  beneficence, 32
  engagement, 34
  harm minimization, 32
  integrity, 32
  justice, 33
  liberty/autonomy, 32
  potential actions, identification of, 30
  privacy/confidentiality, 33
  public benefit, 33
  reflexivity, 33
  relevant ethical merit, 30
  relevant values, identification of, 29–30,
    30*f*
  robust justifications, 31
  transparency, 34
  transparent communication, 31, 31*f*
  trustworthiness, 34

F
Food and Drug Administration (FDA)
  Federal Food, Drug, and Cosmetic Act
    (FDCA), 100
  Medical Device Amendments (MDA),
    98–99
  premarket approval (PMA), 98–99
  requirements for, 100
  two-part test, 98–99

G
Glasgow Index, 45

H
Health inequalities, AI machine learning.
    See Machine learning (ML)
Human-machine interaction
  algorithm aversion, 135–136
  challenges, 135
  randomized clinical trial (RCT), 134–135
  trust building mechanism, 135

I
Image interpretation, 3

L
Label bias, 120
Liability
  apportioning, 93–94

malpractice, 94–98, 95*t*
tort, 91–93

M

Machine learning (ML)
 algorithm
  challenges, 64
  fairness in, 67
  interpretability, 65–66
 category bias, 120–121
 impact of
  health equity, 125–126
  job displacement, 127–128
  social shaping, 126
  surveillance systems, 126–127
  total labor replacement, 127
  World Health Organization (WHO),
   125–126, 125*f*
 interaction with
  healthcare providers, 123
  patients, 123–124
 label bias, 120
 missing data, 122
 model design and training, 122–123, 122*f*
 in model evaluation, 124
 model training problems, 121–122
 prediction, 122
 race/ethnicity, 121–122
 representative bias, 121–122
 social determinants, 125–128, 125*f*
 test-referral bias, 120
Malpractice liability
 challenges, 98
 liability assessment, 97
 medical practice guidelines, 96
 principles, 94–95, 95*t*
 standards of care, 94, 96
Medical artificial intelligence (AI)
 deep learning, 90
 definition, 59
 FDA regulation, 98–100
 liability
  apportioning, 93–94
  malpractice, 94–98, 95*t*
  tort, 91–93
 standards of care, 94–98
ML. *See* Machine learning (ML)

N

National Vaccine Injury Compensation
  Program, 100–101
Natural Language Processing (NLP), 75–76

P

PDSS. *See* Provider decision support systems
  (PDSS)
Personal health information (PHI)
 accountability, 53–54
 biometric data, 46
 challenges of, 46
 confidentiality, 47–50
 consent, 54–55
 data laws, 56
 data protection principles (DPP)
  access and rectification, 53
  accuracy, 51
  data security breaches, 53
  erasure, 51
  explainability, 50–51
  fairness, 50–51
  lawfulness, 50–51
  organizational measures, 53
  purpose limitation and data
   minimization, 51
  retention limitation, 51–52
  right to rectification, 51
  security safeguards, 52
  technical measures, 53
  transparency, 50–51
  use and disclosure, 52
 digital data ethics
  auditability, 57
  ethical considerations, 57
  med-health sciences and services,
   56–57
  profiling, 57
 enhanced privacy rights
  object to processing, 55
  right to data portability, 55–56
  right to restrict, 55
 exemptions from privacy breaches, 50
 Glasgow Index, 45
 law of confidentiality, 45
 personal data/information, 45
 privacy

Personal health information (PHI)
    (Continued)
    crime exemption, 50
    emergency exemption, 50
    in med-health services, 47–48
    Personal Information Protection Law
        (PIPL), 48
    public interest exemption, 50
    right, 46–47
    rule of law exemption, 50
    specific consent, 50
    structure to culture, 57–58
    unintended disclosures, 49
Privacy, 46–50
Procedural ethical issues
    accountability, 34, 41
    engagement, 34, 41
    reflexivity, 33, 40
    transparency, 34, 41
    trustworthiness, 34, 41
Provider decision support systems (PDSS)
    accountability, 41
    autonomy, 40
    beneficence, 39
    engagement, 41
    harm minimization, 39
    integrity, 39–40
    justice, 40
    liberty, 40
    medical futility, 38–39
    public benefit, 40
    reflexivity, 40
    respect for persons, 41–42
    transparency, 41
    trustworthiness, 41

R

Radiomics, 75
Regulation in healthcare
    Advisory Committee on Assisted
        Reproductive Technology
        (ACART), 106–107
    applications, 114
    black box problem, 113
    challenges, 108
    clinical decision support system (CDSS),
        109–110
    consumers' rights, 115
    court judgments, 106
    definition, 106
    guidelines, 107
    healthcare tools, 109
    informed consent, 113
    legal rules, 106
    legislation, 106
    level of risk, 111
    medical device, 109–110
    postmarket, 111–113
    premarket, 111
    smart devices, 110
    software as medical device (SaMD),
        111–112
    Therapeutic Goods Administration
        (TGA), 109–110
    transparency and consent, 113–114
Representative bias, 121–122

S

Science, Health and Policy-relevant
        Ethics in Singapore (SHAPES)
        Initiative, 28
Substantive ethical values
    beneficence, 32, 39
    harm minimization, 32, 39
    integrity, 32, 39–40
    justice, 33, 40
    liberty/autonomy, 32, 40
    privacy/confidentiality, 33, 39–40
    public benefit, 33, 40

T

Test-referral bias, 120
Tort law
    administrative adjudication, 101
    FDA regulation
        Federal Food, Drug, and Cosmetic Act
            (FDCA), 100
        Medical Device Amendments (MDA),
            98–99
        premarket approval (PMA), 98–99
        requirements for, 100
        two-part test, 98–99
    liability
        defective software, 92–93

in diagnosis/treatment, 91
mechanical flaws, 92
negligent credentialing, 92
professional negligence, 91–92
National Vaccine Injury Compensation
    Program, 100–101
no-fault compensation, 101

U
UK Biobank, 17*f*, 18

W
Wearable devices, 21

CPI Antony Rowe
Eastbourne, UK
March 19, 2024